Safety, Systems and People

Safety, Systems and People

Sue Cox BSc MPhil, MIOSH, FRSH
Professor of Health and Safety Management, Centre for Hazard and Risk Management, Loughborough University of Technology, Loughborough, UK

with

Tom Cox BSc, PhD, CPsychol, FBPsS, FRSH, FRSA
Professor of Organizational Psychology, Centre for Organizational Health and Development, Department of Psychology, University of Nottingham, Nottingham, UK

Butterworth-Heinemann
Linacre House, Jordan Hill, Oxford OX2 8DP
A division of Reed Educational and Professional Publishing Ltd

℞ A member of the Reed Elsevier plc group

OXFORD BOSTON JOHANNESBURG
MELBOURNE NEW DELHI SINGAPORE

First published 1996

© Reed Educational and Professional Publishing Ltd 1996

British Library Cataloguing in Publication Data
A catalogue record for this book is available from the British Library
ISBN 0 7506 2089 7

Library of Congress Cataloguing in Publication Data
A catalogue record for this book is available from the Library of Congress

Composition by Genesis Typesetting, Rochester, Kent
Printed and bound in Great Britain by
Hartnolls Limited, Bodmin, Cornwall

Contents

Preface

No society can claim to be civilized if it does not show honest concern for the safety and health of its work-force. Such concern reflects, in part, an understanding of the value of the human resource to the survival and success of its key organizations. This is as true at this point of entry into the age of the knowledge worker as it was during the ascendencies of agricultural and then industrial workers. European and UK employers are charged, in law, with the duty to ensure safe and healthy conditions of work for those they employ. The design and management of such conditions involves more than attention to the immediate physical work environment and work equipment. It involves an understanding and manipulation of the whole 'work system' including aspects such as its culture. The effective management of safety thus begins with an understanding of the nature of the whole system, and of the very nature of systems and of systems management. People are at the centre of such systems but are not its only components. Attention has also to be paid to the technological aspects of those systems. Indeed, the successful management of safety blends knowledge of both of these fundamental resources: the human and the technological. All this, and more, can be achieved within the framework for understanding offered by general systems theory.

This book explores the theoretical underpinning and practical implications of a general systems theory approach to the psychology and management of safety at work. The title – Safety, Systems and People – has been chosen to capture the interplay of the ideas and concerns outlined above, and, at the same time, continue the thesis established by Sue Cox's earlier book with Robin Tait: *Reliability, Safety and Risk Management*.

The information in the present book is presented in a style and at a level appropriate for informed line managers, health and safety and occupational health professionals, final year and Diploma/Masters level students interested in the psychology and management of safety.

Sue Cox is Professor of Health and Safety Management and Director of the Centre for Hazard and Risk Management (CHaRM) at Loughborough University of Technology. Tom Cox is Professor of Organizational Psychology and Director of the Centre for Organizational Health and Development in the Department of Psychology, University of Nottingham, where he is also the current Head of Department. This centre is a World Health Organization (WHO) Collaborating Centre in Occupational Health. The work of both centres is currently funded by both the UK Health and Safety Executive (UK HSE) and the European Commission (DG V), as well as by a wide range of other agencies and organizations. Their approaches to the management of work-related safety and health are complementary, and strongly based in systematic problem solving. Both emphasize the importance of understanding the role that people play in safety and the need to positively manage the human factor. They both exploit systems thinking, and use general systems theory to frame their respective approaches.

General systems theory has been applied to the psychology and management of safety at work for many years. However, the level of sophistication (and understanding) inherent in such application has often been poor and not supported any practical application beyond 'think about it in this way, and see what happens ...'. Furthermore, recent developments in systems thinking are not well represented in such attempts at application, this is particularly true in relation to the development of complexity theory and the theory of organizational health. Both are dealt with in this book, and, in doing so, attention is paid to the particular issue of safety culture.

A system is commonly defined as 'a whole composed of parts in an orderly arrangement according to some scheme or plan' (see, for example, the *Oxford English Dictionary*; also Chapter 4). Such systems are dynamic and can be purposeful. They import things across the boundaries which separate them from their wider environments – energy, information, materials, people – transform them, and then export these outputs back across those boundaries. The concepts of input and output, transformation, and boundaries are important to systems thinking, and these and others are explored in detail in Chapter 4.

General systems theory attempts to explain complex systems and their behaviour. It offers not only a good theoretical framework for understanding the psychology of safety but also the basis for a practical strategy for managing safety. A systems theory based approach is capable of encompassing all our current concerns; from those associated with the more tangible hazards of work and engineering systems, through those associated with management systems, psychosocial hazards and people, to those associated with even more difficult aspects of safety such as safety culture.

Organizational culture has been severally defined, as a concept, by management writers such as Edgar Schein. Whatever, the exact definition with management science, systems thinking treats such culture as an emergent property of the organization as a work system. Each organization encompasses many sub-systems, including the safety management system, and logically each sub-system has its own sub-culture. Thus an organization may have a safety sub-culture. Whether it does or does not, depends on whether there is an obvious safety management system. The overall organizational culture is not necessarily the simple sum of its sub-cultures because the sub-systems, which support these sub-cultures, overlap within the organization. Recent research at Nottingham, by the second author and his colleagues, has allowed a sustainable distinction between the objective and subjective organizations by identifying the sub-systems which comprise the latter. Such research has contributed to the development of a theory of organizational health.

Intertwining the authors' general approach, and its systems theory framework, is a belief in the need to integrate the design and management of safety hardware and software with the management of people. The authors' approach to the challenge of this integration has produced a model which is largely consistent with that subsequently espoused in the 1989 publication by the UK HSE titled 'Human Factors in Industrial Safety'.

The book presents a systems-based view of the psychological and organizational factors and processes underpinning the effective management of safety at work. It is written in three sections, and includes an appendix which illustrates part of the Safety Management Systems (SMS) audit which has been developed at CHaRM, Loughborough University of Technology, based on the ideas presented in the rest of the book. Each of the twelve chapters contain several diagrams to improve the structure of the book and its signposting, and to make it more 'reader-friendly'.

Part 1 presents the background to the book, and begins by asking the question 'why manage safety'? The answer leads into a discussion of the basic equation for safety – that based on the concepts of hazard, harm and risk. This discussion is followed by a review of existing approaches to safety science and safety management and includes a consideration of models of accident causation. Such models, taken together, offer a trace through the history of safety research and, at the same time, an insight into the way such thinking has driven safety management. It is argued that all of the various models have their strengths and weaknesses *when taken separately*, and that what is required is a fully integrated model for understanding and managing safety. However, at the same time, it is also argued that any such model must not relegate its psychological and organizational domains to a secondary position *vis-à-vis* its technical or engineering domains. A general systems theory approach is recommended by the authors, and the various elements of this approach are discussed in Part 2 of the book.

Part 2 is about the psychology of safety, and is constructed around a systems theory model which focuses on the organization, the job (and its constituent tasks) and the person. This approach echoes the recent framework described by the UK HSE for the management of safety – the person in their job (work and work environment) in their organization.

The section begins by offering an explanation and outline of the authors' approach. Central to this approach is the concept of the whole work system. The thinking behind such an approach has been explored in various earlier texts including *Reliability, Safety and Risk Management* by the first author and Robin Tait (1991), and 'Psychosocial and Organizational Hazards: Control and Monitoring' by both authors (1993). This sets up a discussion of general systems theory applied to organizations and safety. This discussion takes thinking in this area up to recent developments including complexity theory. Throughout the emphasis is on the interaction between the three components of the model: the organization, the job (and its tasks) and person. The authors then consider each of those components in turn beginning with the organization. Their discussion of the organization explores the general issues of organizational structure and function, and organizational behaviour in relation to safety. It also looks at the currently 'hot' issue of organizational culture. It does so with a healthy degree of scepticism and attempts to anchor its discussion in empirical studies (something of a rarity in this area). This introduces the second author's (T Cox) work on organizational health. The final chapter concerns the topic of work-related stress.

Part 3 focuses on the application of the information introduced in the preceding three sections: the management of safety. It is the nub of the text for the practitioner and for managers and policy makers. However, having said that, none of these groups should consider taking this section in isolation from those that precede it.

Part 3 begins by explicating the authors' framework for managing safety which derives directly from their overall model. It sets this discussion within a second framework – a strategy for risk assessment – risk management – as described in current UK and EC legislation on safety. It considers two contrasting meta-strategies for managing safety at work: the organizationally-focused and the individually-focused. The first is presented in two stages: an opening discussion of management systems and procedures with the clear objective of 'safe management' and, then, a discussion of 'safe systems of work' and 'safe workplaces'. The discussion of individually-focused strategies looks at the different ways in which the individual worker's knowledge, skills and attitudes might be developed towards the notion of the 'safe worker'. The concept of the 'safe worker' is explored, and with those of 'safe management' and 'safe systems of work' offered as another equation for thinking about the objectives of managing the whole work system.

The book is concluded with a brief résumé of its main thesis, and some thoughts, from the authors, for the future of safety research and safety management.

The authors hope that this book will prove both interesting and informative, and that it will encourage and inform good safety management not only in the UK but also in other European countries and further afield. Certainly, the first author's experiences in the Peoples Republic of China and in Japan, and in Kuwait have led her to believe that the approach described here, although developed and fine tuned in studies in west Europe, is more widely applicable. It remains a substantial but potentially manageable challenge to adapt the approach to the challenges not only of the Middle and Far East but also to those of east and south Europe and the former Soviet Union. Whatever, systems thinking should offer a powerful tool for unlocking the secrets of good safety management, and enhance our understanding of what underpins that management. Safety is all about systems and people: safety, systems and people.

Acknowledgements

The authors gratefully acknowledge the help and support of Sandy Edwards and Alistair Cheyne. Without their unstinting efforts, this book would never have seen the light of day. They would also thank their other colleagues for their support, not only while the book was being completed but over the years leading up to this project when the idea behind it was being formed and developed.

The authors also gratefully acknowledge the support, both tangible and intangible, of the UK Health and Safety Executive, the Building Research Establishment (Garston, Watford), the World Health Organization (Regional Office for Europe), the European Commission, and the International Labour Office. Most of all they wish to acknowledge the support and encouragement of their respective Universities. The views expressed here are those of the authors and do not necessarily reflect those of any other persons or organizations.

Part 1
Background

This first part of the book considers three questions, the answers to which provide the essential platform on which it can then explore the psychology and management of safety. These questions concern first, the need to manage safety, second, the nature of hazard, harm and risk, and the relationships between them which provide the basic equation underpinning our understanding of safety, and finally, the nature of safety science and its attempts to conceptualize accident causation.

The three chapters in Part 1 are:

1 Why manage safety?
2 Hazard, harm and risk: the basic equation
3 Safety science and safety management

Part 1
Background

1
Why manage safety?

Introduction

Successful organizations actively manage all aspects of their business including safety and, for such organizations, safety is of equal importance to production (or service provision), quality and environmental control. Indeed those organizations that are capable of managing safety well also tend to perform well in the other domains (Health and Safety Executive (HSE), 1981). There is evidence to support observations that the skills required for the management of safety are similar to those required to ensure excellence in both quality and production (HSE, 1991; Petersen, 1971). At the same time, it is obvious that there is a real cost associated with not managing safety (see later).

Although many managers and safety practitioners might agree in principle with the views expressed in the previous paragraph, they would not necessarily agree that safety management is a priority within their own organization. They may rationalize this apparent contradiction by expressing the belief that:

1 safety management is only a necessity in particular industries, for example mining, oil extraction, nuclear or chemical processing;
2 safety management is only important within large businesses and not within small or medium sized enterprises (SMEs);
3 only certain activities within organizations, for example, production processes rather than service activities, are important management domains for safety;
4 safety is the responsibility of individual workers and is not a line management responsibility; or, finally,
5 accidents and incidents can be considered in isolation of the organizational context in which they occur.

In the past, many of these beliefs have been reinforced by decisions taken by the insurance industry. Historically, the most 'costly' hazards, in terms of insurance claims, existed in the manufacturing industry and were associated with the activities of manual workers (Sun Alliance International, 1991). Furthermore, the focus of insurance inspections has traditionally been on the physical environment rather than on organizational and management systems. However, recent settlements for upper limb disorders in office workers (Medical Law Report 129, 1991) and for occupational stress in social work (T Cox et al., 1995) are challenging insurers to review their approach. As a consequence all work activities and all aspects of work environments are being more closely scrutinized for potential health and safety liabilities. In many cases, organizations have experienced increased financial outlay for employee liability insurance (Davies and Teasdale, 1994).

The negotiations around premium setting have prompted many organizations to review their policies and systems for the management of safety. However, even when an organization accepts the need to manage safety, it may still perceive that such a function is outside the remit of directors and line management. Organizations that subscribe to this point of view tend to relegate safety to a subsidiary function – one which may be devoid of an adequate power base. They may also ascribe poor safety performance to the actions and misactions of their work-force and use traditional models of accident causation to explain the relationship between employee behaviour and accidents (see Chapter 3).

This chapter explores the nature of work and work environments in the 1990s, attempts an overview of associated safety problems and, in doing so, presents the case for adopting a proactive approach to safety. It argues that, despite the legal framework for the regulation of safety within the United Kingdom (UK) and Europe, accidents and work-related ill-health still pose considerable problems for many organizations. It further argues that such problems could be well enough managed by a systematic problem solving approach such as that outlined in the final sections of this book.

Work and the work environments: the source of risks

Types of work and work environments are extremely diverse and their designs are largely dependent on the nature of their primary work activities. For the majority, work is supported by specific tools, machinery and related facilities (hardware), formal systems and procedures (software) and the social context of their organization. The effective design of work and work environments recognizes the importance of all three (S Cox et al., 1995). However, the quality of work performance is dependent not only on effective design and on employee behaviour but also on the demands of external agents such as 'customers', shareholders, financiers, politicians, etc.

These agents are part of the wider environment and together the various elements referred to in this paragraph make up the total 'work system' (see Figure 1.1 and Chapter 4).

Forecasters of change in the nature of work (see, for example, Toffler, 1981) have heralded the emergence of the post industrial society in which the focus, organization and status of work have changed. The *focus* has moved away from person centred production to the use of automated and computer based technologies. The *organization* of work has moved away from complex hierarchical structures to smaller and devolved structures and

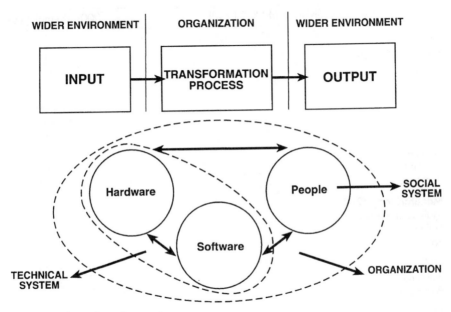

Figure 1.1 The total 'work system'

to 'virtual' organizations. The *status* of work in society has changed so that the centrality of work and the pattern of career development are considerably different for the worker in the 1990s than they were for his or her counterpart in the 1960s. Such changes are occurring and are having considerable impact on the attitudes, motivations and behaviour of today's work-force in relation to safety. At the same time, while the emergence of new forms of work and work environments may eradicate traditional work hazards, for example, entrapment in moving machinery, new work-related hazards, including psychosocial hazards (see Chapter 2), may be created. In relation to safety, changes in the nature of work may at one and the same time solve old problems and generate new ones.

The population at 'risk'

Evidence on current numbers and types of employees is collected through statistical surveys by the Department of Employment. Information is available from the 1991 Census of Employment for Great Britain and the UK (Employment Gazette, 1993) which describe the distribution and patterns of

Table 1.1 Distribution and patterns of workers from the 1991 census of employment for Great Britain and the United Kingdom (UK)

Industries and services (divisions of SIC[1] 1980)	000s of employees
All (0–9)	
Male and Female	22 111
Full-time	16 358
Part-time	5 753
Male	11 448
Full-time	10 433
Part-time	1015
Female	10 661
Full-time	5926
Part-time	4735
Manufacturing (2–4)	
Male and female	4678
Full-time	4335
Part-time	343
Male	3279
Full-time	3223
Part-time	56
Female	1399
Full-time	1112
Part-time	287
Services (6–9)	
Male and female	15 695
Full-time	10 448
Part-time	5247
Male	6736
Full-time	5836
Part-time	900
Female	8960
Full-time	4613
Part-time	4347

[1]Standard Industry Classification

work (full-time or part-time). It is summarized in Table 1.1. The 1991 census collected information on those in employment during September of that year. It covered organizations of all sizes throughout all sectors of the economy. The following trends have emerged since September 1989 (the date of the previous census):

1 over 22 000 000 workers were in paid employment within Great Britain and the UK in September 1991; however,
2 the total number of employees has decreased by three per cent;
3 more employees were employed within the service sector but fewer in the manufacturing sector; and
4 there was an increase in part-time employment, particularly amongst female workers.

Today's workforce (according to the 1991 Census) is therefore more likely to work in environments which utilize information rather than manufacturing technology and to work in environments which have traditionally been perceived to be 'low risk' and are, in theory, less likely to suffer serious injury or harm from work accidents. Much of their work appears to take place within offices and commercial rather than manufacturing premises. Against traditional beliefs about the 'safeness' of work and work environments, workers should be less at risk now than they were in the past. The next section discusses the major sources of statistical data for workplace safety which both highlight the 'size' of the current problem and provides an insight into the nature of safety risks.

The nature of the risk

Safety statistics are available from several key sources, including the:

1 UK HSE and Local Authority databases;
2 survey data gathered by Government departments (for example, the supplement to the Labour Force Survey, 1990);
3 organizational records; and
4 specific research projects.

While the UK HSE and similar data provide an overview of the number and nature of accidents (and incidents of work-related ill-health) across work environments, organizational records provide evidence of accident patterns within particular companies and institutions.

UK HSE data

The Statistical Unit of the UK HSE interprets the data that are reported to the Executive under statutory requirements of Reporting of Industrial Injuries,

Diseases and Dangerous Occurrence Regulations 1985 (RIDDOR). These statistics are summarized in a statistical supplement to the Health and Safety Commission's (HSC) Annual Report (see, for example, HSC, 1994). One of the recent supplements contained provisional 1992/93 statistics as well as the final figures for 1991/92. The following data (based on 1992/93 preliminary statistics) are abstracted from this supplement (see Table 1.2).

Table 1.2 Injury incidence rates (per 100 000 employees) by industry, 1992/3 (HSC, 1994)

Industry	Fatal accident	Major injury accident	Over-three-day injury accident
Agriculture, forestry and fishing	7.8	165	558
Energy and water supply industries	4.9	194	1640
Manufacturing	1.4	125	1120
Construction	7.5	241	1368
Services	0.7	53	479
All industry	1.3	81	677

Table 1.2 describes estimated injury incidence rates (per 100 000 employees) by industrial sector. These figures reflect the relative risks faced by workers within agriculture, forestry and fisheries, energy and water supply, manufacturing, construction and service industries. The data may also be used to examine the nature of the accidents which occur (see, for example, Table 1.4 adapted from S Cox and O'Sullivan, 1995) and of the injuries experienced by workers by sectors.

The data are based on statutory reporting requirements and are not necessarily comprehensive since there is no requirement to report injury accidents which do not result in at least three days' absence from work. Comparison of these data with evidence from the 1990 Labour Force Survey (LFS) suggests that a significant number of non-fatal injuries are under-reported. There are a number of reasons why such under-reporting occurs, not least of which may be ignorance of the law. For example, some employers bring employees back to work on 'light' duties so as to avoid reporting thresholds. Despite the problem of under-reporting, and taken as a whole, the HSE data reveal a number of interesting findings, including:

1 combining all sectors of industry, workers have approximately a 1 in 150 risk of experiencing an accident which results in losing over three days' absence from work;

2 this risk is greater for workers in the construction and energy and water supply industries than in the other sectors; and
3 the major injury incidence rate in some sectors is 'unacceptably' high, for example, agriculture (etc.), construction and energy and water (etc.).

Furthermore, although service sector 'environments' are perceived to be low risk, fatal accidents still occur in this sector and workers still experience injuries as a direct consequence of workplace accidents.

Labour Force Survey

Additional data on the number of accidents by sector have been published by the UK HSE (Stevens, 1992) following their trailer to the 1990 LFS. Table 1.3 shows the number of injury accidents reported in the LFS supplement. These data are based on structured interviews with members of about 600 000 private households throughout Great Britain during March, April and May 1990 (that is, about one in every 350 private households in Great Britain). Respondents were asked about workplace accidents occurring within the past twelve months, and work-related illness they had suffered during the past year (whether linked to work in the course of the year or as a consequence of past work). The results have been scaled to give estimates

Table 1.3 Injury numbers by industry (England and Wales) (adapted from LFS, 1990)

Industry	SIC classification	Number of injuries	Injury rate per 100 000 workers
Agriculture	0	37 776	8141
Energy and water	1	39 673	8213
Extraction of minerals	2	64 125	8460
Metal goods	3	179 034	7658
Other manufacturing	4	147 059	6856
Construction	5	198 719	10 859
Distribution, hotel and catering	6	249 874	5226
Transport	7	97 365	6633
Banking, finance	8	58 664	2157
Other services	9	348 694	5420
Not known/missing		3253	
Total		1 424 236	6081

relating to the whole population in private households in Great Britain in spring 1990 (Stevens, 1992).

The total number of injuries (Davies and Teasdale, 1994) is believed to reflect an underestimate of the total problem because:

1 they include only the most recent work-related injury suffered by an individual and do not allow for more than one injury per person in a year;
2 the numbers reported for earlier months are successively less than the latest three months, which may reflect ease of 'recall';
3 the LFS questionnaire could be answered by proxy; and
4 the LFS is household based so it does not collect information from people in communal establishments (for example, nurses and members of the armed forces).

Davies and Teasdale (1994) have constructed an accident triangle (see Figure 1.2) based on the LFS survey. This triangle was used to illustrate the fact that, for every fatal injury accident, there were 280 major injury accidents, 1320 over-three-day injury accidents and 1975 minor personal injury accidents. A cursory examination of these figures would suggest that the ratio of minor injury to over-three-day injury accidents is lower than that reported in other accident triangles (see, for example, Tye, 1976). This may reflect the biases of the LFS study discussed above.

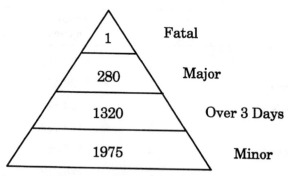

Figure 1.2 Accident triangle illustrating LFS Data (adapted from Davies and Teasdale, 1994)

Figure 1.2 reports the relative risk of being killed to sustaining a minor injury as 1 in 1975. However, before any action could be taken to reduce this risk, a more detailed analysis of the nature of those accidents would be required. The data would need to be explored further in the particular context of interest. This is illustrated by the first author's research on safety in buildings (S Cox and O'Sullivan, 1995).

The nature of accidents

Research on safety in buildings (see S Cox and O'Sullivan, 1995) has identified the most common causes of accidents associated with the design and use of all buildings (including domestic and leisure buildings). These accidents have been elaborated using a specially developed accident taxonomy (see Table 1.4). The taxonomy is based on eight primary accident categories. Each of the accidents described in the taxonomy may result from exposure to a plethora of hazardous events (see S Cox and O'Sullivan, 1995, for a full description of this work).

One of the most common causes of accidents associated with the use of buildings is slips, trips, and falls (Woodall and S Cox, 1993). Their aetiology may be described in terms of three factors: first, the physical environment, the floor surface or stairway on which the incident occurred; second, the ambient environment including lighting, humidity, etc.; and finally, the characteristics of the individual (or victim) to whom the accident occurred. Individual accidents occur as a result of a unique combination of all three factors in this 'system' which involve a consideration of person–job–organization factors (see Chapter 4). However, detailed examination of a significant number of accidents within this category may reveal particular trends and allow a further description of hazardous events.

Risk assessments should incorporate this type of analysis in the estimation and control of hazards (see Chapter 2). The supporting evidence on the relative contribution of each of these factors to safety outcomes within buildings may also be derived from statistical data. Understanding the nature of accidents in this detail is necessary in order to reduce the associated risks. Understanding the extent of the problem is a necessary first step in estimating its costs.

The costs of workplace accidents

Managers in industry and commerce know that accidents cost money but often experience difficulty in estimating the size of that cost. The direct costs to an organization can be measured in terms of lost time, damage to work equipment and premises, productivity losses, and costs of replacement goods and services. The indirect costs are more difficult to estimate. Some are of a secondary nature, for example, replacement costs for injured workers. Some are hidden, for example, costs arising from loss of commitment and motivation of workers involved in accidents. Some are delayed, for example, the loss of good workers as a result of post accident trauma. In many studies, estimates of these indirect costs have been included (see, for example, Morgan and Davies, 1981).

Table 1.4 Building related accident taxonomy (adapted from Building Regulation and Safety, S Cox and O'Sullivan, 1995)

Category of accident		
i	Slips, trips and falls (STF) in and around buildings	movement or elevation between different levels of buildings leading to STF
		movement on the same level flooring in a noisy environment and poor lighting conditions leading to STF
		movement on the same level on the curtilage or outside pathways of buildings leading to STF
		falling out of the window or balcony
		falling down steps
		falling on uneven surfaces
ii	Collision with parts of buildings	impact with doors without glass
		impact with doors with glass
		impact with ceiling in stairwells
		impact with glass
		impact with fixed installations
		impact with building fixtures in conditions of poor lighting
iii	Entrapment in part of building	entrapment in doors
		entrapment in windows
		entrapment in escalators/lifts
		entrapment in heating fixtures
iv	Hit by falling objects	tiles falling from roofs and hitting person(s)
		cladding falling from building and hitting person(s)
		collapse of part of building (for example, ceiling)
		collapse of outbuildings
		catastrophic collapse
		scaffolding falling during maintenance
v	Explosion	faulty gas installations and appliances (gas safety)
		pressurized vessels explode
vi	Electrical accidents	contact with electrical current
		contact with fixed electrical installations
		lightning striking person(s)
vii	Burns and scalds	contact with lights, heaters, etc.
		contact with hot water
viii	Poisoning/asphyxiation	contact with poisonous gas

A study on the overall cost of accidents and work-related ill-health to the British economy (Davies and Teasdale, 1994) has revealed the following:

1 the total estimated annual cost to the economy is between £11 and £16 billion (i.e. a thousand millions);
2 these estimates are equivalent to between 2 per cent and 3 per cent of the Gross Domestic Product;
3 the cost to employers is between £4 and £9 billion annually which is equivalent to between 5 per cent and 10 per cent of all UK companies' gross trading profits in 1990;
4 these costs are equivalent to £170 to £360 for each person employed;
5 the cost to individual workers and their families is estimated at nearly £5 billion (after making allowance for social security benefits and compensation payments).

One of the conclusions which was drawn by authors of this report was that managers do not fully appreciate the costs of accidents to their business. In the same vein, the Chairman of the HSC, Frank Davies, has been quoted as saying: 'It isn't more regulations we need but it is an appreciation that accidents will hurt your balance sheet'.

Major disasters such as those arising from major fires or explosions or involving loss of life are clearly associated by employers with large financial losses. For example, the Piper Alpha explosion involved the loss of 167 lives and is estimated to have cost over £2 billion including £746 million in direct insurance payouts. Equally, the mechanism for compensating workers is fairly well understood by them: it is increasingly 'formula-driven' (Kemp, 1988). Less well understood, however, is the mechanism for costing accidents and incidents of a more routine nature. As a result, a UK HSE report, 'The costs of accidents at work' (HSE, 1993), has described a series of case studies in accident costing. The key findings of this study are detailed in the following section.

UK HSE accident costing studies

The case studies in accident costing were carried out by the UK HSE's Accident Prevention Advisory Unit (APAU) (HSE, 1981). Five case studies were carried out within different sectors of industry including offshore oil extraction, construction, dairy and food manufacturing. Studies were also carried out in the public sector within the hospital environment. The common aim underpinning these studies was the development of a methodology and supporting procedures to accurately identify the full cost of accidents.

The basis of the APAU methodology was loss control theory (Bird and Loftus, 1974). The methodology was based upon an extremely broad definition of the term 'accident'. For the purpose of the study an accident was defined as 'any unplanned event that resulted in injury or ill-health of people, or damage or loss to property, plant, materials or the environment or a loss of business opportunity'. Data were gathered on all accidents meeting this definition (and involving loss over a previously agreed threshold) over a period of several months. Pro forma were used in the data collection and these have been published by the UK HSE (1993). The cost of each accident was then assessed and a judgement was made on cost effective prevention methods. These accident costs were, *inter alia*, expressed in relation to other operational costs, for example;

1 accidents cost one organization as much as 37 per cent of its annual profits;
2 a second organization incurred accident losses equivalent to 8.5 per cent of the tender price of a major contract; and
3 accident costs to a third organization were about 5 per cent of its running costs.

None of the participating organizations suffered major or catastrophic losses during the study periods nor were there any fatal injuries. Despite this, the results illustrated the fact that accident costs could be significant to the financial viability of the organization.

Although the APAU studies demonstrated some of the real costs of accidents, the methodology used was (by necessity) complex and the UK HSE was not able to produce a 'readily applied and simple formula' envisaged by organizations such as the Confederation of British Industry (CBI). Not withstanding this criticism, the exercise was valuable, and the costs identified were consistent with those associated with non-compliance in quality (Whiston and Eddershaw, 1989). Further information on the methodology used in the case studies has been published by the HSE (1993) and can be obtained from the APAU in Bootle, UK.

Finally, the APAU case studies also showed that an estimated 10 per cent of all accidents could have resulted in more serious injuries. This is a significant finding and one which argues for an approach in which all accidents, including minor incidents, are systematically investigated. For example, a worker may trip over and drop a pile of files while walking up to an upstairs archive room. The consequence may be a 'near miss incident' – all the files land on the stairs. Alternatively, the worker may hurt him- or herself and sustain minor injuries or, in the worst possible case, he or she may fall backwards and suffer serious injury or death. The occurrence and the extent of the injury may largely be a matter of chance. The lesson is clear; all incidents need to be prevented, even ones which appear comparatively minor.

So far the authors have presented economic arguments in favour of managing health and safety. There is, however, another and possibly even more compelling argument for taking appropriate actions, that is the fact that the law demands it.

The legal requirements to manage health and safety: risk regulation

The responsibility for ensuring the health and safety of employees rests primarily with the employer. These duties extend to all types of 'workplace' and 'work activity' (Handy and Ford, 1993). They are the basis of Section 2 of the Health and Safety at Work etc. Act 1974 (HASAWA), which is the focus of UK legislation on health and safety. Furthermore, these duties have been reinforced and elaborated by European legislation and, in particular, the Framework Directive (number 89/391/EEC) on 'measures to encourage improvements in the health and safety of workers' which is now implemented through the Management of Health and Safety at Work Regulations 1992 and the associated sets of regulations. These are considered in more detail in Part 3 of this book.

Managerial responsibility for safety

In recent years there have been a series of major disasters in work systems which have highlighted the role of managerial control in safety. For example, the Seveso accident (1976), the Tenerife air disaster in 1977, the Three Mile Island two years later, Bhopal (1984) with its considerable loss of life, Chernobyl (1986) with its implications for the public image of the nuclear power generating industry, and the Piper Alpha oil platform explosion (1988) have all received media coverage. The public enquiries (see, for example, Cullen, 1990) have pointed out managerial and organizational failures in the aetiology of these accidents. Management responsibility for safety is specified in HASAWA under Section 37 of the Act (HASAWA, 1974) and there are a number of successful prosecutions taken under this section every year.

Yet despite this possibility for prosecutions, there is a growing lobby for the law to recognize managerial responsibility for an organization's safety performance. For example, a private bill was drawn up by David McIntosh (1991) recommending that safety obligations be included within directors duties and James (1992) has suggested that regulations should be introduced under Section 235 of the Companies Act 1985. His recommendations include:

1 a tightening up of accountability for senior managers in relation to their management decisions and their impact on safety; and
2 an introduction of some form of no fault liability for the director named as responsible for safety.

The enactment of more legislation may have considerable impact on managerial decision making. On the one hand, fear of being 'scapegoated', should an accident occur, may result in improved decision making or alternatively, it may result in uncertainty and a reluctance to make safety-related decisions (this is further discussed in Chapter 9).

Breaches in health and safety legislation

Employers and their managers who are in breach of their statutory duties are liable to criminal prosecution and may incur substantial fines (see above). They may also incur liability in civil law (Munkman, 1990) for any damages suffered by employees. Injury claims can amount to a substantial sum when considered at the organizational level (Davies and Teasdale, 1994). The increases in employee liability insurance premiums which have tripled over the past five years (HSE, 1991, 1993) are indicative of such costs. The adverse publicity following both the incident and the subsequent prosecution may also have a significant effect on the corporate image and thus affect business opportunities. There are several examples of organizations who have lost market share as a direct result of a safety-related incident (including the parent organization of the Piper Alpha).

Organizations and their social context: the risk environment

The third set of reasons for serious investment in managing safety relate to the wider social context in which business is being conducted today. Businesses are being increasingly challenged to meet demands for social responsibility. Pressure groups and lobbyists for safe practices no longer reflect idiosyncratic views of radical minorities but they echo the views of customers and consumers. The accepted norms for effective production advocate that businesses should achieve their goals without harming employees. Safety is, thus, a social and a moral issue. The fact that an increasing number of organizations report their performance in Annual Reports confirms the importance of safety to shareholders and investors. Their interest reflects the symbiotic relationship between social responsibility and overall profits. Organizations cannot afford to ignore health and safety requirements because safety is not only a legal responsibility, it is a business imperative. It is at the very core of corporate values and reflects its ethos and its goals.

Summary

This chapter has considered work and work environments in the 1990s and has reviewed statistical and survey evidence on the nature of safety 'risks'.

It has further argued that such risks should be both 'systematically' and 'actively' managed within all work organizations and has justified these activities in relation to the following arguments:

1 organizations and their managers have legal duties to manage safety;
2 safety is also a social and a moral issue and organizations are increasingly being challenged to meet their social responsibilities; and
3 safety is good business (the financial imperatives).

The legal, moral and cost dimensions thus provide the motivators for action. The following chapters in this section of the book further explore our understanding of safety risks as a basis for subsequent discussions.

References

Bird, F.E. and Loftus, R.G. (1974). *Loss Control Management*. Loganville: Institute Publishing.

1991 Census of Employment for Great Britain (1993). *Employment Gazette*, April.

Cox, S. and O'Sullivan, E. (1995). *Buildings Regulation and Safety*. London: Construction Research Publications Ltd.

Cox, S., Janes, W., Walker, D. and Wenham, D. (1995). *Office Health and Safety Handbook*. London: Tolley Publishing Company.

Cox, T.R., Griffiths, A.G. and Stokes, A. (1995). 'Work-related stress and the law: the current position'. *Journal of Employment Law and Practice*, **2**, 93–96.

Cullen, D.W. (1990). *The Public Inquiry into the Piper Alpha Disaster*. Department of Energy, HMSO.

Davies, N.V. and Teasdale, P. (1994). *The Costs to the British Economy of Work Accidents and Work-Related Ill-Health*. Sudbury, Suffolk: HSE Books.

Employment Gazette (1993). 1991 Census of Employment for Great Britain. *Employment Gazette*, **101**(4).

Framework Directive 89/391/EEC (1989). *The Minimum Requirements for Protecting the Health and Safety of Workers*. Brussels, 12 June 1989.

Handy, J. and Ford, M. (eds) (1993). *Redgraves, Fife and Machin: Health and Safety* (2nd edn). London: Butterworths.

Health and Safety at Work etc. Act 1974. London: HMSO.

Health and Safety Commission (1985). *Reporting of Industrial Injuries, Diseases and Dangerous Occurrences Regulations (RIDDOR)*. London: HMSO.

Health and Safety Commission (1992). *The Management of Health and Safety at Work Regulations 1992*. London: HMSO.

Health and Safety Commission (1994). *Annual Report 1992/1993* (1994). London: HMSO.

Health and Safety Executive (1981). *Managing Safety: A Review of the Role of Management in Occupational Health and Safety by the Accident Prevention Advisory Unit of HM Factory Inspectorate*. London: HMSO.

Health and Safety Executive (1991). *Successful Health and Safety Management*. London: HMSO.

Health and Safety Executive (1993). *The Costs of Accidents at Work (HS(G)96)*. London: HMSO.

James, P. (1992). 'Reforming British health and safety law: a framework for discussion'. *Industrial Law Journal*, **21**(2), 83–105.

Kemp, D.A.M. (1988). *The Quantum of Damages in Personal Injury and Fatal Accident Claims*. London: Sweet and Maxwell.

McIntosh, D. (1991). 'Private Member's Bill'. *New Law Journal*, October 25.

Medical Law Report 129 (1991).

Munkman, J. (1990). *Employers' Liability* (11th edition). London: Butterworths.

Morgan, P. and Davies, N. (1981). 'Costs of occupational accidents and diseases in Great Britain'. *Employment Gazette*, November, **89**(11), 477–485.

Petersen, D. (1971). *Techniques of Safety Management*. New York: McGraw-Hill.

Stevens, G. (1992). 'Workplace injury: a view from HSE's trailer to the 1990 Labour Force Survey'. *Employment Gazette*, December, **100**(12).

Sun Alliance International (1991) *Gradually developing diseases: A guide*. London: Sun Alliance International.

Toffler, A. (1981). *The Third Wave*. London: Pan Books Ltd.

Tye, J. (1976). *Accident Rate Study 1974/75*. London: British Safety Council.

Whiston, J. and Eddershaw, B. (1989). 'Quality and safety – distant cousins or close relatives?' *The Chemical Engineer*, June, 97–102.

Woodall, F.J. and Cox, S.J. (1993). 'Stumbling into a serious accident'. *Health and Safety Practitioner*, May, 42–46.

2
Hazard, harm and risk: the basic equation

Introduction

The concept of absolute safety is a myth. There is no such thing as a 'totally' safe work environment, system of work or work activity; all are unsafe to some degree and are characterized by risk profiles which are organization and job specific. Individuals always face some risk (however small) of being harmed wherever they are and whatever they are doing: at work, at home or while engaged in leisure. In the case of fatal accidents, the probabilities are very low indeed. In 1991 in Britain, with a population of 57 million, there were 12 816 accidental fatalities out of a total of 628 000 deaths from all causes (Office of Population Censuses and Surveys (OPCS), 1991). Of these, less than 500 were work-related.

Many definitions of safety introduce the notion of 'acceptable' (or 'unacceptable') risk and link the concepts of risk and harm. For example, Fido and Wood, (1989) define 'safety' as a state of 'freedom from unacceptable risk of personal harm' (see Chapter 3). This chapter examines the nature of the overall hazard–harm–risk process which represents the basic equation underpinning the safety management process in relation to work. It is structured in five sections:

1 the first section considers the notion of a 'hazard' and explores how it is currently used in framework legislation on health and safety and in management strategies;
2 the second section examines the nature and severity of 'harm' and describes the authors' work on the scaling of harms;
3 the third section describes the nature of the pathways linking hazards and harm;
4 the fourth section considers the notion of risk; and finally,
5 the last section outlines the processes underpinning risk assessment.

The final part of the chapter draws heavily on the 'Risk Assessment Toolkit' (S Cox, 1992) published by Loughborough University of Technology. It also introduces the nature of organizational and managerial decision making for risk which is further elaborated in Part 3 of the book.

Hazard

A hazard is defined as an event or situation which carries the potential for harm (T Cox and S Cox, 1993). Work hazards can be broadly divided into the more tangible and *physical*, which include the biological, biomechanical, chemical and radiological, and the *psychosocial*. The concept of a psychosocial hazard may be difficult to grasp. Such hazards relate to the interactions among job content, work organization and management, and environmental and organizational conditions, on the one hand, and the workers' competencies and needs on the other (International Labour Office (ILO), 1986). Psychosocial hazards have gained enhanced recognition as a result of recent cases in law related to occupational stress (see, for example, T Cox *et al.*, 1995).

However, this simple model of hazard should be developed in a number of ways (T Cox and S Cox, 1993). First, it is necessary to introduce two further and related concepts: the 'hazardous situation' and the 'hazardous event' (S Cox, 1992). A hazardous situation is a situation (or set of circumstances) in which a person interacts with the hazard but is not necessarily exposed to it. A hazardous event is the trigger which exposes the person to the hazard; it initiates the chain of events leading to harm. The hazard can be an aspect of work or of the work system, organization or environment; the hazardous situation is effectively that aspect in use. For situations involving acute exposure (see later), the hazardous event often describes the breakdown of use – the error or accident.

Two examples may serve to illustrate these points (see Table 2.1). The first example is taken from research into risk perception in nurses being conducted by the Centre for Organizational Health and Development at the University of Nottingham and funded by the United Kingdom Health and Safety Executive (UK HSE) (T Cox *et al.*, 1993). The second example is taken from research into the safety of buildings (S Cox and O'Sullivan, 1995) funded by the United Kingdom Building Research Establishment (UK BRE).

The first example concerns nurses' exposure to a particular microbiological hazard. In this example (Table 2.1), the actual hazard and agent of harm is the human immunodeficiency virus (HIV). The hazardous situation for nurses is nursing contact with patients who are HIV positive, and the hazardous event (error, accident or technical failure) may be a needlestick injury which results in exposure to the patient's infected blood. HIV carries a potential for harm, but it is only when the nurse works with that hazard, through contact with an infected patient, that such harm may be expressed (hazardous situation) and only then when a breakdown in safe working practice occurs (hazardous event). The person is at risk of harm in a hazardous situation. The sequence of events leading to that harm is triggered by the hazardous event.

The second example relates to slips, trips and falls on stairs (see Woodall and S Cox, 1993). In this example (Table 2.1), the hazard or agent of harm is the stairs. The hazardous situation for the person as the stair-user is ascending or descending those stairs, and the hazardous event (error or accident) is slipping or tripping. Stairs carry a potential for harm, but it is only when the person uses the stairs that such harm may be expressed (hazardous situation or hazard in use), and only then when a breakdown in

Table 2.1 Hazard, hazardous situation and hazardous event: concepts defined

Concept	Example 1	Example 2
Hazard Aspect or characteristic of work or work environments	HIV	Stairs
Hazardous situation Working with hazard (or hazard in use)	Nursing HIV-infected patients	Person ascending or descending stairs
Hazardous event Accident, error or technical failure	Needle stick injury causing contact with infected blood	Slipping, tripping on stairs, leading to falling

system safety occurs. It is interesting to note that such a breakdown may not only relate to the person's actions or those of others but may be caused, for example, by inadequate design or maintenance of stair treads, etc.

The second way in which the simple model of hazards needs to be developed (T Cox and S Cox, 1993) relates to a distinction between acute and chronic exposure. Both of the examples described in Table 2.1 are based on acute rather than chronic exposure. However, if an example of chronic exposure had been used (for example, low level exposure to organic lead) then that exposure to the hazard would be ongoing. In this case there would have been a convergence between the notions of the hazardous situation and the hazardous event. The hazardous situation would represent in itself a slowly forming accident or 'slow accident'.

Exposure to many psychosocial hazards of work is chronic rather than acute. Arguably, the most obvious exception is exposure to traumatic events, such as those constituting workplace violence. The problems associated with assembly line work, for example quantitative work overload, qualitative work underload and lack of control over the pacing of work, are chronic in nature and offer a contrast to those associated with violent incidents. Working on the assembly line can represent both the hazardous situation as well as the hazardous event: it can be the archetypal 'slow accident'. The harm associated with exposure to such work has been reviewed elsewhere (T Cox, 1985).

Types of hazard

As already suggested, work hazards can be broadly divided into the more tangible and physical, which include the biological, biomechanical (and ergonomic), chemical (and fire) and radiological, and the psychosocial. Such hazards may also be grouped by the legislative demands on employers. For example, while legislation exists on the control of noise, asbestos, hazardous substances, etc. (see Handy and Ford, 1993), the control of other hazards, many of them psychosocial, is not covered by specific legislation but comes under the auspices of the 'framework' legislation (see later).

Physical hazards
A wide variety of more tangible and physical hazards has been extensively studied for their effects on workers' health. Most can be measured objectively, and with some degree of reliability and validity, and they are therefore relatively easily monitored in the workplace (see, for example, S Cox and Tait, 1991; Ridley, 1993). The evidence suggests that such hazards represent significant challenges to both physical and psychological health. In some cases, standards exist which can be used in the regulation of exposure to these potential sources of harm.

Psychosocial hazards
Psychosocial hazards are those which relate to the interactions among job content, work organization, management systems, environmental and organizational conditions, on the one hand, and workers' competencies and needs on the other. Those interactions which prove hazardous influence workers' health through their perceptions and experience (ILO, 1986). This point is explained more fully in Chapter 8. Evidence suggests that certain psychosocial characteristics of work are associated with the experience of stress, job dissatisfaction and ill-health (see T Cox, 1993).

Hazard identification

Hazard identification is one of the primary processes of safety management. The three main categories of method used for identifying 'hazards', including hazardous situations and hazardous events, are intuitive, inductive or deductive (see Table 2.2 adapted from S Cox, 1992).

Intuitive techniques are used to rely and build on the person's report of their experience of the hazards in their immediate work environment. They provide a useful starting point for any subsequent analysis and are particularly useful when consensus can be reached at the level of the work team (Walker and S Cox, 1995). Inductive techniques focus on the 'what if' question and offer a useful way of elaborating on the hazardous situation. Finally, deductive techniques explore hazardous events (or accidents) in a logical and structured way to see how those could (or did) happen. For example, Figure 3.5 (see next chapter) is the result of a deductive investigation and describes a fault tree for the overturn of a container in a nuclear facility. Each of these methods of hazard identification often include

Table 2.2 Methods of identifying 'hazards' (adapted from S Cox, 1992)

Method	Example
Intuitive	Brainstorming
Inductive 'What could go wrong?' 'What if. . .?'	Failure mode and effect analysis Hazard and operability study Analysis of potential problems Action error analysis Job Safety Analysis Key points/checklists
Deductive 'How can it happen?'	Fault tree technique Accident analysis

techniques which are used in other areas of business activity (for example, quality, production or environmental management). A full description of these techniques is outside the scope of this particular book. What is important here is that the reader appreciates the breadth and scope of hazard identification methods. The interested reader is referred to other texts (for example, Andrews, 1993; S Cox and Tait, 1991; Kletz, 1971; Thomson, 1987).

Harm

Harm (to an individual) may be defined as the adverse effect on that individual which may arise from exposure to a hazard; it is the damage done to the individual or the loss they sustain. There are various categories or types of 'individual' harm (see Table 2.3) and many can have both short and long term effects. To a certain extent, these categories may only be considered as broad and indistinct representations of potential harms (Marshall, 1987). They are not rigidly segregated from one another, nor is the boundary between short term 'harm' or long term 'harm' strictly defined. Furthermore, any particular harm might be caused by exposure to one or more of a variety of hazards. At the same time, a single hazardous event may result in several types of harm (S Cox and Tait, 1991). If we refer back to one of our earlier examples (see Table 2.1), a building user may slip and fall on the stairs for one or more of a number of reasons and either be killed, suffer a minor injury or longer term emotional trauma, or escape unharmed. Equally, that person might have been killed by one of a whole range of different incidents in the building.

Table 2.3 Categories of 'individual' harm arising from hazardous events (adapted from S Cox and Tait, 1991)

Death	Immediate or delayed
Physical injuries	Disabling and non-disabling
Disease	Immediate or delayed
Mutagenic effects	Short or long term
Teratogenic effects	Immediate or delayed
Psychological 'injuries'	Short and long term
Social trauma	Short and long term

The different categories of harm vary in their nature and individual harms vary in their severity, both within and between categories. The consequences (to the individual) of the hazardous event define the severity of the harm and represent the reasons why, or ways in which, it damages the person's health and quality of life and alters their normal behaviour.

Measurement of harm

The measurement of harm can take several different forms, depending on the type of harm under consideration, for example, general wellbeing, disease and injury. It has been argued that an ordinal scale of injury is by far the easier harm classification to both understand and apply (Soby and Ball, 1991). An example is provided by the Abbreviated Injury Scale (see Yates,

Table 2.4 Examples of rated harms from different outcome categories[1] (adapted from Cheyne et al., 1994)

Class	Original Scale	Psychological/ Behavioral	Musculo-skeletal	Respiratory
V	2.00 to 3.49	Occasional slight restriction of social behaviour	Temporary slight backache	Occasional mild cough/cold
IV	3.50 to 4.49	Occasional severe discomfort	Moderate ligament injury in hip	Occasional mild pneumonia Regular serious cough/cold
III	4.50 to 5.49	Chronic immediate severe stress	Temporary paralysis of arms	Occasional moderate pneumonia
II	5.50 to 6.49	Severe chronic confusion/ dementia	Chronic severe backache	Regular severe fever
I	6.50 to 7.00	Suicide	Permanent paralysis below neck	Regular severe pneumonia

[1]The full range of harms are published in Raw et al., 1995

1990). However, this scale purposely makes no attempt to cover broad health outcomes and focuses exclusively on physical injuries.

The authors (Cheyne et al., 1994) have developed a rating scale which measures the severity of all potential harms which may befall building users. For an example of part of the scale, see Table 2.4 (adapted from Cheyne et al., 1994). The development took place under four discrete stages.

1 During Stage 1 experts in building hazards (including medical and safety practitioners, fire experts and health physicists) were guided through a structured brainstorming exercise and generated a comprehensive list of harms; these harms covered all the categories outlined in Table 2.3 earlier.
2 During Stage 2 the list of identified harms was rationalized by the researchers to produce a final list of approximately three hundred separate harms. The basis of the rationalization process was first, to ensure that the list was internally consistent at the level of description,

and second, that the harm terminology could be understood by people outside the medical profession. Finally, redundant items were removed.

3 During Stage 3 two approaches to the development of a measure of severity were explored using the final list. The first approach involved the overall comparison of the different harms and direct rating of the severity of individual harms to construct an ordinal scale. The second approach involved identification of the various dimensions of consequence for the harms and then the rating of the individual harms against scales developed to reflect those dimensions of consequences.

4 During Stage 4 the two resulting scales were compared. This comparison demonstrated an overall level of agreement between the two different scales (despite some differences in the lower and mid points of the scale). This result supported the hypothesis that there is a single underlying factor of severity of harm and that a single scale can be used (Cheyne *et al.*, 1994). This was confirmed by factor analysis of the consequence data.

A more complex approach to the measurement of wellbeing is illustrated in the development of the health index (Teeling Smith, 1985). The index combines scores obtained from measurements of disability and discomfort to form a score indicating current health state. While this is a sound approach for the measurement of the wellbeing of individual patients (Teeling Smith, 1985), it may not be appropriate in measuring harms in large populations of workers and in a variety of hazardous situations. Soby and Ball (1991) argue, for example, that clinically developed disease-specific indices are likely to be incompatible with aggregated statistical data (see Chapter 1), mainly due to person and situational specific dimensions (see Torrance *et al.*, 1982, for an example of such an index).

Relationships between hazards and harm

It has already been established (see earlier sections) that exposure to physical and psychosocial hazards may threaten both physical and psychological health and that such harm is measurable. The evidence suggest that such effects may be mediated by, at least, two pathways (see Figure 2.1): first, a direct physicochemical mechanism as, for example, in the effects of dust inhalation as a contributory factor in pneumoconiosis; and second, a psychophysiological stress-mediated mechanism as, for example, in the effects of perceived loss of control as a possible contributory factor in coronary heart disease. Both may have behavioural correlates. They do not offer alternative explanations of the hazard–harm–health relationship and, in most hazardous situations, *both* operate and interact to varying extents and in various ways (Levi, 1984; T Cox and S Cox, 1993). Levi (1984) has

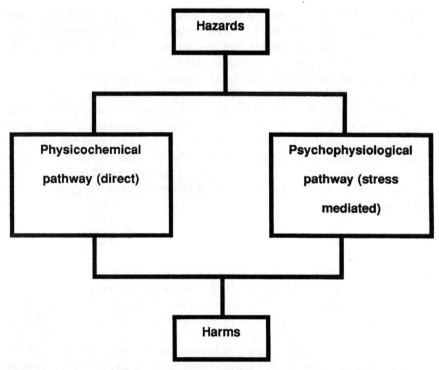

Figure 2.1 Pathways from hazards to harms (adapted from T Cox and S Cox, 1993)

noted that both additive and synergistic interactions[1] are possible. Examples of such interactions may exist in relation to work-related upper limb disorders (Griffiths *et al.*, 1996) or the alleged reproductive health effects of exposure to visual display units (HSE, 1992).

Many of the effects of psychosocial hazards are undoubtedly mediated by psychophysiological – stress-related – processes (see Chapter 8), while many of the direct physicochemical effects of work on health are accounted for by more tangible physical hazards. Despite this, physical hazards can affect health through psychophysiological pathways (Levi, 1984), while certain psychosocial hazards may have direct effects. For example, exposure to organic solvents may have a psychological effect on the person through their direct effects on the brain, through the unpleasantness of their smell and

---

[1] The outcome of effects which interact additively is simply the sum of the separate effects; however, the outcome of effects which interact synergistically may be different from the sum of the separate effects. It may be greater, where one set of effects facilitates or enhances another, or it may be smaller, where one set attenuates or weakens another.

through fear that such exposure might be harmful (Levi, 1981). The latter can give rise to the experience of stress. Violence, as a psychosocial hazard, may have a direct physical effect on its victim, in addition to any psychological trauma or social distress that it causes.

Acceptance of the basic principle underpinning this argument takes us beyond 'equivalence reasoning' (T Cox and S Cox, 1993); that is, only expressing concern for the direct physicochemical actions of the more tangible physical hazards or for the psychophysiological actions of psychosocial hazards. The model offered of the relationships between hazards and harms is, therefore, that the different types of hazard can each contribute to the various forms of harm. Furthermore, significant inter-actions can occur both in the interplay of hazards and in the patterning of their health effects.

Risk

Risk has been variously defined in the available literature. The report of the Study Group of the Royal Society (1983) defines risk as 'the probability that a particular adverse event occurs during a stated period of time, or results from a particular challenge'. As a probabilistic measure, risk must always relate to a certain defined hazardous event or set of hazardous events and, where appropriate, should relate to a known level of exposure (see earlier). The definition also implies that harm results from the event. The concept of risk is thus used in various ways to quantify the relationship between hazards and harm and to provide some measure of the likely harmful effects of hazards (see Figure 2.2).

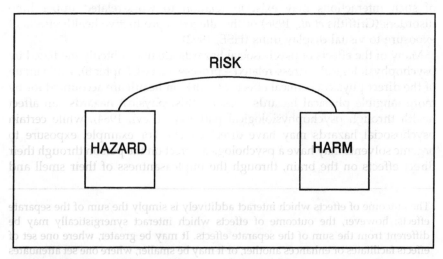

Figure 2.2 The concept of 'risk' as a link between hazard and harm

Risk equations

The two main dimensions of 'risk' (see equation 1) that may be used in its quantification (or estimation) are:

1 probability or likelihood of harm occurring in relation to a particular 'hazard'; and
2 magnitude or severity of harm.

The probability of harm occurring is often assumed to be some function (f) of exposure to the hazardous situation (equation 2) which, in turn, can be defined as the product of frequency of exposure and duration of exposure (equation 3). An approximation to exposure is often made in terms of the occurrence of the hazardous situation (see Warner, 1992). However, measures of occurrence of hazardous situations and those of exposure to the hazard are not identical. At the same time, the process which takes the person from exposure to the hazard to the defined state of harm may involve a number of stages. Each stage may have an associated probability of occurring, given completion of the previous stage. For example, a nurse suffering a needlestick injury involving HIV infected blood may or may not pick up that infection and, if they do, may or may not (in a given time) develop AIDS. The probabilities of HIV infection and the development of AIDS are different, and the latter subsumes the former.

The following set of equations based on this rationale have been published by the authors (T Cox and S Cox, 1993).

Equation 1: Risk = f (probability of the defined harm, magnitude or severity of that harm)

Equation 2: Probability of the defined harm = f (exposure to hazardous situation)

Equation 3: Exposure to hazardous situation = f (frequency × duration)

Bringing these equations together, risk can therefore be defined as a function of the product of the frequency and duration of exposure to the hazard, and the magnitude or severity of the defined harm (equation 4).

Equation 4: Risk = f (frequency ×duration of exposure to hazardous situation, magnitude or severity of the defined harm)

Hazards have been described in various ways (see earlier). One important dimension is that of exposure: some hazards are obviously more usually associated with acute exposure while others are commonly associated with more prolonged or chronic exposure. Acute exposure to the hazard usually occurs as a result of human error or technical failure in an otherwise safe system and can be characterized more or less as an 'off–on' switch. In the chronic situation, the hazard, whatever its cause, has been present for some time and the person experiences continual exposure in an essentially unhealthy system. This situation can be characterized more or less as an 'on' state. For these situations, the acute and chronic poles of the exposure dimension, the risk equation can be rewritten as follows.

For the acute situation, frequency of exposure to the hazard may be more important than duration, thus the equation tends to be:

Equation 5: Risk = f (frequency of exposure to the hazard,
                           magnitude or severity of the defined harm).

For the chronic situation, duration of exposure to the hazard may be more important than frequency, thus the equation tends to be:

Equation 6: Risk = f (duration of exposure to the hazard,
                           magnitude or severity of the defined harm).

The definition of risk in mathematical terms supports a natural science approach to the quantification of risk through the use of available empirical data. Risk measures derived in this manner are deemed to be objective (S Cox and Tait, 1991). However, there are at least two other related aspects of the risk concept that need to be considered in practice:

1 attributable and relative risk, as a comparison between the likelihood of harm attributable to exposure to different hazards; and
2 acceptable or tolerable risk (see later), as the level of risk that can be accepted or tolerated for those exposed to the hazard.

Interestingly, the nature of the risk function (see above) may vary in relation to people's perceptions with the type of hazard being considered. Although it is frequently argued that the function is essentially multiplicative, some have argued that it is additive (Kaplan and Garrick, 1981). Is the risk function dependent on the product or the sum of the hazard exposure and severity of harm terms? Recent evidence from studies on nurses' anxiety about microbiological hazards (T Cox et al., 1993) has suggested that the type of function which operates depends on the nature of the hazard involved. For more common microbiological hazards, where knowledge of the hazard–harm relationship is good, a multiplicative function appears to exist, while for less common microbiological hazards, where knowledge is poorer, then an additive function provides the better explanation. Here the size of the 'dread' factor (see Slovic *et al.*, 1981) relating to the magnitude or severity of harm *drives* the perceived risk equation.

Measuring risk

Risk can be measured using a variety of techniques and at differing levels of precision and reliability. Measurement relies on some degree of quantification. The more accurate and reliable the measure is then the more useful it is. Many contrast objective and subjective measures of risk without offering a clear statement of what is included in those two broad categories. Measures of risk based on reliability, counting publicly verifiable events are clearly objective. Those based, for example, on self-reports of feelings about situations, such as anxiety, are clearly subjective, although their reliability can be established and their validity could be explored. What, however, is the status of a person's estimate of the frequency of occurrence of publicly verifiable events, where the reliability and accuracy of such estimates can be established? Is this risk perception? The answers are not clear in much of the literature and a more careful terminology is required. Distinctions are required between objective risk measures, subjective estimates of risk, perceptions of risk in terms of subjective evaluations (such as those described by Slovic *et al.*, 1981), and the emotional correlates of risk, such as anxiety. Risk perception is about how people evaluate risks and it is essentially a process of cognitive appraisal – a process which imbues situations with personal meaning (Holroyd and Lazarus, 1982). It is different from, but relates to, subjective estimates of objective risk and its emotional sequel, etc.

Most expressions of risk relate the likelihood of a defined harm from a specified 'hazard' to population levels. For example, approximately 4500 people in England and Wales die as a result of home accidents (S Cox and O'Sullivan, 1995). If the total population in England and Wales is 48 million (Central Statistical Office, 1988), the risk of death due to an accident in the home is approximately 1 in 10 000 per annum. Similar expressions of fatal risks can be worked out for all the 12 816 accidental deaths mentioned in the introduction and for specific workplaces. However, for each fatal incident there are countless potentially fatal events (Adams, 1995). The problem for safety professionals and decision makers is that the relationship between potentially fatal events and actual fatalities may become extremely tenuous as a result of risk compensation (van der Schaaf *et al.*, 1991). Because events are recognized as potentially fatal, people take more care. Equally, if hazardous situations are perceived as 'safe' people may behave in an unsafe manner (Howarth, 1987).

In the workplace people have a much greater chance of sustaining minor injuries than being seriously injured or killed (see Chapter 1). However, the data on minor workplace injuries is notoriously unreliable (Stevens, 1992).

The challenge for management is to assess *all* workplace risks to assure effective control. The process of risk assessment is described below.

Risk assessment

Risk assessment is concerned with the prevention of accidents and ill-health in the workplace. It is a proactive process which involves the analysis and subsequent evaluation of risk (see Figure 2.3). Risk assessment is now formally required as part of the duties of employers (HSC, 1992) as the cornerstone of their safety management system. Figure 2.3 (adapted from S Cox and Tait, 1991) acknowledges the overall importance of risk assessment and places it in the even broader context of organizational risk management.

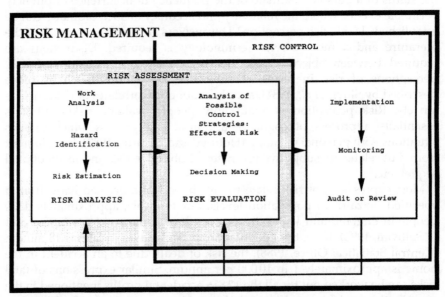

Figure 2.3 Risk assessment as part of the overall process of managing risks (adapted from S Cox and Tait, 1991)

The requirement for an all embracing risk assessment is contained in Regulation 3 of the Management of Health and Safety at Work Regulations 1992 (HSC, 1992), which requires every employer, and in some instances the self employed, to make a suitable and sufficient assessment of:

1 the risks to the health and safety of employees to which they are exposed whilst they are at work; and

2 the risks to the health and safety of persons not in his employment but arising out of or in connection with the conduct by him or his undertaking.

In addition to this wide ranging requirement to carry out risk assessment on all hazards, there are a number of regulations covering specific hazards (HSE, 1994) which require prescribed assessments.

There are a number of established methods for assessing risks (see, for example, AIChemE, 1989; Booth, 1993; S Cox, 1992; Engineering Employers Federation (EEF), 1993; Kazer, 1993). These involve qualitative and/or quantitative methods (see, for example, HSE, 1989; S Cox and Tait, 1991; Thomson, 1987). However, all such methods require a comprehensive analysis of potential risks, together with some acceptable outcome measures. The next sections describe risk analysis and risk estimation and consider their importance in the overall risk assessment process. Finally, the applications of the Risk Assessment Toolkit (S Cox, 1992) are discussed.

Risk analysis

Risk analysis can be either a qualitative or quantitative process (Andrews, 1993), partly depending on the nature of the risk. For example, quantitative risk analysis is a technique which has been used extensively within the aerospace, electronics and nuclear industries and which is now being employed in many other industries (albeit in a modified way) as a direct consequence of recent legal requirements for risk assessment (see above). Risk analysis is used to quantify (or model) the likelihood of either a specific incident or event or a sequence of events.

The American Institute of Chemical Engineers (AIChemE, 1989) provides a good description of the methodology underpinning risk analysis in the process industry. There are essentially seven stages in its implementation. These are as follows.

1 *System description*, which is the compilation of all technical and human information needed for the analysis.
2 *Hazard identification*, which is a critical step. A hazard omitted at this stage is a hazard which is not subsequently analysed or considered.
3 *Incident enumeration*, which is the identification and tabulation of all hazardous incidents (or events) without regard to their importance or to the initiating event.

Stages (2) and (3) may be linked. For example, while chlorine gas is identified as a 'hazard', its unplanned emission through a faulty valve is the 'hazardous incident'. Similarly, HIV is identified as a 'hazard' (see earlier), while a needlestick injury causing contact with infected blood is the 'hazardous event'. Hazard analysis and operability studies (HAZOP) and other methods are used in hazard identification and incident enumeration (see earlier).

4 *Incident frequency estimation*, which uses likelihood estimation models for selected incidents and evaluates frequencies. Fault Tree and Event Tree Analysis (S Cox and Tait, 1991) and the Technique for Human Error Rate Prediction, THERP (Humphreys, 1988) are typical techniques used at this stage (see Chapter 6).
5 *Consequence estimation*, which is the methodology used to determine the potential for damage or harm from specific incidents.
6 *Evaluation of consequences*; this stage is concerned with the estimation of frequency data for specified consequences. Estimates can be based on existing databanks (see, for example, HAZDATA, 1990 and Boff and Lincoln, 1988) and also on other sources of historical data, including organizational accident and incident data.
7 *Risk estimation*, combines the consequences and likelihood of all incident outcomes from all selected incidents to provide a measure of risk.

One of the authors (S Cox, 1992) has described a risk matrix approach to risk estimation (see Figure 2.4) in which likelihood and severity of harm (consequence) are plotted on the matrix in order to compare separate risks. The matrix offers nine possible risk profiles ranging from high likelihood/ high severity through low likelihood/low severity. It has been used to support risk evaluation (see below).

All these seven stages in the AIChemE (1989) methodology have been discussed in greater detail in S Cox and Tait (1991).

Risk evaluation

Rational decisions about the acceptability (or tolerability) of risks should only be made on the basis of criteria which are usable, transparent and agreed by all the parties concerned. Within an organization interested parties would include workers who are exposed to the risks, managers who have management responsibility for the viability of the work system and technical assessors who may wish to see higher safety standards in place. Equally, they may be parties external to the organization, regulators, shareholders or members of the public (see Chapter 1). There is also a need to ensure that such criteria are suitable for comparisons to be made of risks of different harm outcomes, for example minor and fatal injuries, and acute or chronic health effects.

Experts in nuclear engineering (HSE, 1988) have proposed tolerable limits of safety in large scale industrial organizations on the basis of what has been observed to be accepted, or at least tolerated, by those exposed. The

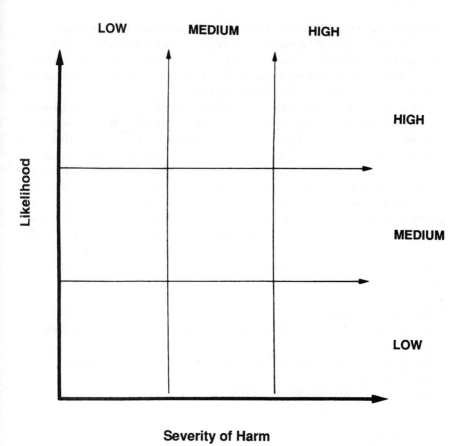

Figure 2.4 The risk matrix (adapted from S Cox, 1992)

following quote from 'The Tolerability of Risk from Nuclear Power Stations' (HSE, 1988) states:

> it seems right to suggest that the maximum level that we should be prepared to tolerate for any individual member of the public from any large scale industrial hazard should not be less than 1 in 10 000 (1 in 10^4) [per year]. Such a level would as it happens equate to the average annual risk of dying in a traffic accident, and can be compared with everyone's general chance of contracting fatal cancer which is an average of 1 in 300 per annum (HSE, 1988).

The single figure tolerable limit of fatal injury of 1 in 10^4 per annum is clearly usable in the particular context of fatalities, comprehensible to scientists, and transparent (it is not easy to hide dead bodies). However, the figure cannot be used directly to assess the tolerability of non-fatal injuries or illness and may not be fully understood by all the workforce. Moreover, third parties, such as affected members of the public, may not welcome exposure to any risk if no benefit accrues to them. Risk evaluation is a complex rather than a straightforward process.

The evaluation of any risk is dependent on the person's (or group's) perception and knowledge of that hazard and the associated consequence, and in turn their experience of that or similar hazards, together with perceived benefits. The role of cognitive processes in the evaluation of risk means that its outcome may be different in kind or degree from the quantified risk analysis discussed above (see Chapter 7). The evaluation of risk is first dependent on who the 'assessor' is (Hale, 1986). The overall process of risk assessment may also be fundamentally different when applied to individual and social behaviour than when applied to the behaviour of engineering systems (Covello, 1983, 1989). Hale (1986) has reviewed research into subjective evaluation of risk and has highlighted additional factors including:

1 the nature of the hazard;
2 the extent to which exposure to the hazard and its potential for harm are controllable;
3 the timescale over which harm may occur (i.e. acute versus chronic harms);
4 the assessor's knowledge and understanding; and
5 the magnitude of the imagined consequences.

Human behaviour is by and large not solely determined by the 'objective' estimation of risk as calculated by numerical methods (see earlier). In some cases the objective measurement of risk will match the person's subjective estimation and indeed may have played some part in determining their perceptions and emotional reactions. Interestingly, the person can 'come into' the process of risk analysis at any of its many stages (see Figure 2.3), and thus be given anything from a vague feeling of dread and anxiety (pre-hazard identification) to a subjective risk estimate (post-estimation) to work with. Furthermore human behaviour towards danger cannot be explained or predicted using a single measure of harm (see Chapter 7). Beyond this the collective perception of the person's reference group (the social perception) may also strongly determine their individual view (see Douglas and Wildavsky, 1982; Douglas, 1992) as, at an even higher level, may the public perception.

Often, public policy is determined more by collective perceptions of risk than by its more objective estimation (Petts, 1989). For example, the resources devoted to industrial safety far exceed those dedicated to road or home safety, yet compared to the latter the workplace is a relatively safe environment. There are about 450 deaths per annum in workplaces in the UK, while some 4500 are killed on the roads each year, and 6000 are killed through accidents in the home (in the UK as opposed to England and Wales) in the same period. Public reaction to safety is also more related to people's perceptions than to mere 'objective' fact. For example, public reaction to road deaths, occurring in many different accidents each involving a relatively small number of people, is very different to their reaction to air crashes. Although the latter involve more people per event, the total number killed per year is far fewer than on the roads. It would require say twenty jumbo jet crashes each year in Britain to match the death toll on our roads. Obviously to understand what is happening in anecdotal examples such as these, the perception of risk has to be systematically studied in relation to a particular work activity and to the population who may be 'at risk'.

'Acceptability' or 'tolerability' of risk

Judgements on the 'acceptability' or 'tolerability' of risk of harm from a particular work activity may be made using the principle of reasonable practicability. The UK HSE (HSE, 1988; O'Riordan, 1985) has developed this in the ALARP principle (see Figure 2.5).

ALARP principle

The ALARP principle states that risks must be made *As Low As Reasonably Practicable* and, unless the expense required to avoid harming either the workforce or the general public is in gross disproportion to the risk, the employer must undertake that expense.

The interpretation of gross disproportion will vary with the level of risk under review. The UK HSE (HSE, 1989) has adopted a three-tier risk standard. According to this approach, depicted in Figure 2.5, very high levels of risk are totally unacceptable and must be reduced even at very high cost. By contrast, for very small risks it is not normally required that further significant expenditure be committed to achieve even higher levels of safety. These are considered acceptable or tolerable risks. In between the very small and the very large risks lies a wide range of risk levels to which the ALARP principle applies. Table 2.5 illustrates the HSE (1988) objective criteria to support this principle.

The decision to go ahead with risk-reducing expenditure for each project or activity involving risk levels in this middle range, usually follows a judgement by the particular decision-maker within an organization and these decisions are further regulated by judgement on the part of the UK HSE inspectors. Some of the main factors used by the UK HSE in judging whether a risk level is above, below or within this middle range of risk levels include:

1 what the annual risk of death for each age group is against that for all other causes;
2 how the risk being considered compares to other risks known to be 'acceptable';
3 whether the risk affects those on site (i.e. workers) or off site (i.e. the public); and
4 whether children, the elderly, or disabled persons are at risk.

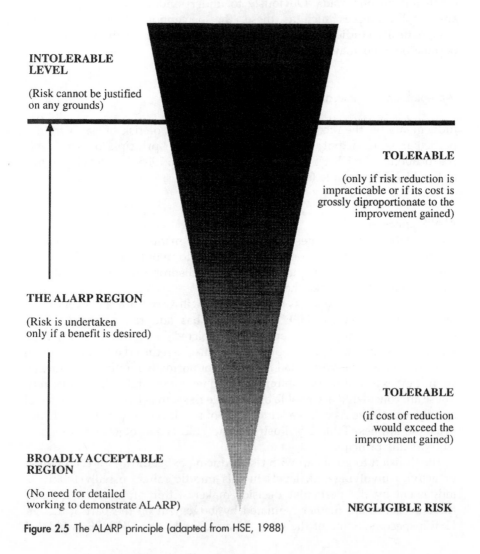

INTOLERABLE LEVEL

(Risk cannot be justified on any grounds)

TOLERABLE

(only if risk reduction is impracticable or if its cost is grossly diproportionate to the improvement gained)

THE ALARP REGION

(Risk is undertaken only if a benefit is desired)

TOLERABLE

(if cost of reduction would exceed the improvement gained)

BROADLY ACCEPTABLE REGION

(No need for detailed working to demonstrate ALARP)

NEGLIGIBLE RISK

Figure 2.5 The ALARP principle (adapted from HSE, 1988)

Table 2.5 Tolerable and acceptable risk criteria (HSE, 1988)

	Chance of death per annum
Maximum tolerable risk for employees	1 in 10^3
Maximum tolerable risk for members of the public from any large-scale industrial hazard	1 in 10^4
Broadly acceptable risk: the risk below which it would not be reasonable to insist on expensive further improvements in standards	1 in 10^6

Organizational considerations

The organizational considerations involved in the evaluation of risk will vary from organization to organization, and from hazardous situation to situation. However, the main issues will probably include:

1 the general acceptance criteria used within the organization; these criteria require clear operating standards, appropriate arrangements and management systems;
2 the results of a cost–benefit analysis: costs of implementing controls against the reduction of risks which may accrue;
3 the identification of humanitarian issues: injuries and ill-health cause real pain and suffering; and
4 the recognition of legislative constraints (or opportunities): organizational standards should at least meet legislative requirements.

The end result of risk evaluation could be a number of action priority levels (the organizational equivalent of 'ALARP'). These levels will be negotiated between the risk assessment team and senior management when the risk assessment system is being devised, and depend on the organizations' resources. One approach (S Cox et al., 1995) may be to develop three action levels: immediate action, action within a specified time and no action, but monitoring to ensure the situation does not change. Table 2.6 illustrates a possible link between the risk matrix (see Figure 2.4) estimates with suggested action levels. These action priority levels have been used in work environments to support decision making for risk (see S Cox et al., 1995).

The Risk Assessment Toolkit

Previous sections of this chapter have examined the nature of the hazard–harm–risk relationship and have outlined the concepts underpinning risk

Table 2.6 Action priority levels in support of the risk matrix

Category	Grid reference	Action priority
A	High likelihood/ high severity	Immediate action. High likelihood of serious injury.
B	High likelihood/ medium severity	Immediate action. Reduce likelihood.
C	High likelihood/ low severity	Seek longer term means of reducing likelihood. May need to judge priority.
D	Medium likelihood/ high severity	Plan reduction of likelihood of event. Consider design of lower severity system. Judge priority.
E	Medium likelihood/ medium severity	Plan reduction of likelihood of event. Consider design of lower severity system. Judge priority.
F	Medium likelihood/ low severity	Judge priority. Long term plan to reduce likelihood.
G	Low likelihood/ high severity	Monitor standards to reduce/maintain likelihood to lowest possible level. Seek design of lower severity system as priority.
H	Low likelihood/ medium severity	Monitor to maintain standards. Consider the possibility of lower severity system.
I	Low likelihood/ low severity	Monitor annually to ensure likelihood does not increase.

assessment. This section introduces a simple five-step approach, based on the Risk Assessment Toolkit (S Cox, 1992), which is easily adapted to meet the assessment needs of most types of organizations. (Readers are advised to refer to the source document for a more in-depth discussion.)

The key steps in the toolkit process are outlined in Figure 2.6 which flow-charts the process.

The Risk Assessment Toolkit, therefore, includes not only guidance on the analysis of risks, but also includes advice on the evaluation of different control options and can thus be considered an integral part of the safety management process (see Chapter 9). This evaluation builds on the existing systems and practices for controlling hazards within the workplace and is fundamentally about choosing appropriate control strategies.

Some of the key steps are considered below.

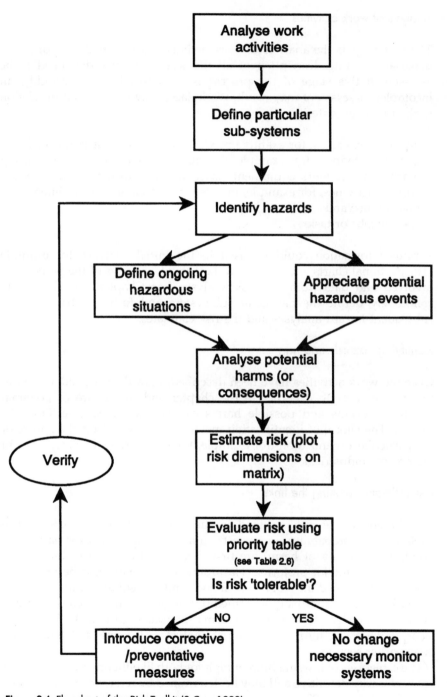

Figure 2.6 Flowchart of the Risk Toolkit (S Cox, 1992)

Analysis of work activities

The first stage is the analysis of work activities. It is initially important to make an accurate description of all the work activities that need to be assessed. If this stage of the process is not carried out thoroughly, an incomplete assessment may result. Workplaces may be analysed in relation to the following criteria:

1 geographical areas, for example production facilities, car parks, etc;
2 generic activities, for example the manual handling of loads, using machinery and work equipment, using display screen equipment, etc;
3 specific activities, for example, maintenance of plant and machinery, fire precautions; and
4 specific jobs or tasks.

Various information could be used in the initial analysis, for example organizational charts, interviews, 'walk through' surveys of the work areas involved and data from job/task analyses (see Chapter 11). These data collectively define the particular sub-system and form the basis of any subsequent hazard analyses and the risk estimates.

Identifying hazards and harms

Once the work activities have been described, hazards can be identified in the manner described earlier in the chapter and the nature of potential hazardous events and possible harms can be elaborated (see Table 2.2 earlier). The choice of identification methods usually reflects the nature of the particular organization (for example, HAZOPs are extensively used in the process industries).

Risk estimates: plotting the links

The dimensions of risk outlined in the risk equations (see earlier: the likelihood (or probability) of harm from a specified hazard (or hazardous event) and severity of harm) are first determined and are systematically recorded on a toolkit proforma. Hazardous events may then be ranked in terms of low/medium/high likelihood and potential harms may be similarly ranked in terms of low/medium/high severity (or consequence). They are then plotted on the risk matrix (see Figure 2.4). This stage incorporates stages four to seven of risk analysis (AIChemE, 1989) discussed in an earlier section.

The categories of low, medium or high for both likelihood and severity are not necessarily definitive and may be determined by reference to assessment grids (see, S Cox, 1992). A more complex approach may be adopted to suit specific organizational needs. For instance, a four by four matrix will clearly allow more categories as there will be sixteen choices as opposed to the nine choices described by Figure 2.4 and Table 2.6. Some organizations may prefer to use a numerical approach (Steel, 1990) to risk estimation. Risks may be expressed numerically as the product of the likelihood (on a scale of one to five) and the consequence (determined on a similar scale). This approach is outlined in the Health and Safety Commission (HSC) publication on the management of health and safety at work (HSC, 1992). The toolkit (S Cox, 1992) describes a matrix-based approach and has been used in a variety of workplaces including offices and manufacturing facilities (Walker and S Cox, 1995). It is also the basis of a computer-based risk assessment system (see Chapter 9).

Risk evaluation and control options

Risks are evaluated according to organizational criteria which incorporate the considerations discussed earlier. For example, the organization will usually have clearly defined operational standards which conform with existing legislative frameworks. They may also have policies with respect to safety issues (see Chapter 9) which require that employees are not harmed in the course of their job. Having said this, most organizations have some room for improvement in safety (see Chapter 1) and the evaluation of risks leads them to debate the effectiveness of their existing controls. Table 2.6 illustrates possible control options for particular risk 'plots'.

For each hazard identified, therefore, the final stage of the risk assessment process thus involves the consideration of:

1 existing controls, and whether they are effective;
2 the possible implementation of new/refined control measures; and
3 the priority of implementation, based on the risk evaluation.

All these decisions need to be considered in the context of broader decision making issues which face managers in all types of organization.

Organizational (and management) decision making for safety

Organizational decision making theories (see, for example, Harrison, 1987) often assume a rational problem solving approach. For safety this might be described as:

1 identifying the problem (in this instance the nature of the workplace risk);

2 generating alternative solutions (in this case the consideration of 'plausible' and 'acceptable' control strategies);
3 evaluating and choosing solutions;
4 implementing the chosen solutions to control the risk; and finally
5 monitoring, reviewing and appraising the solution.

These are the steps in the risk assessment–risk management approach illustrated in Figure 2.3.

However, this ideal process ignores the fact that within an organization different 'stakeholders' will inevitably have different perspectives on the problem (for example, safety representatives and safety advisers may view safety-related risks differently from line managers). Similarly, some alternative solutions will not be tolerated within the organization, for example, those which appear to restrict profit generating may be considered as 'non-starters'. Thus if the organization has asked a group of people to make effective decisions as part of the risk assessment process (see Figure 2.3) then difficulties could occur. First the group may not reach a consensus and second this may lead to a revision of the decision making cycle.

Glendon and McKenna (1995) have described a 'cycle of decision failure' which may be typical of many safety decisions made by groups. In this cycle (Glendon and McKenna, 1995) decision making may be described as:

1 a problem arises;
2 a solution is proposed (before options are considered);
3 the decision process is orientated towards this solution;
4 further issues emerge (these are ignored);
5 commitment to the initial solution is increased so as to eradicate the problem;
6 more resources (including time, money and people) are committed to this initial solution;
7 failure occurs; and
8 new problem(s) emerge.

In a study (Wilson and Rosenfeld, 1990) of 150 strategic decisions within organizations, 60 per cent were observed to follow the decision failures cycle described above. Roberts and his co-workers have reviewed organizational behaviour in the light of decision making theories and believe that such theories assume 'organizations can absorb errors with few negative consequences' (Roberts et al., 1994).

However, Reason's (1990) theory of accident development (see next chapter) identifies faulty decisions at all levels of the organization as the root cause of failures that lead to accidents. Clearly decision making for safety is an important part of effective risk control and is considered in later chapters.

Summary

This chapter has examined the nature of the overall hazard–harm–risk process which represents the basic equation underpinning the safety management process in relation to work. It has considered the notions of 'hazard', 'harm' and 'risk' separately and then considered the process of risk assessment as a basis for legal compliance in relation to safety. It is clear that safety management cannot be discussed without an understanding of the basic equations outlined in this chapter. Together the concepts of 'hazard', 'harm' and 'risk' provide the language of safety management. On first acquaintance they appear to be relatively straightforward. However, all three have to be used consistently and sensibly to assure safe work. 'Hazard', 'harm' and 'risk' need to be carefully defined and explored within the particular organizational contexts and decision making on 'risk' needs to be both rational and logical. This chapter has attempted to address these needs and also provides a necessary background and introduction to the safety science and approaches to accident causation addressed in the next chapter.

References

Adams, J. (1995). *Risk*. London: University College Press.

American Institute of Chemical Engineers (AIChemE) (1989). *Guidelines for Chemical Process Quantitative Risk Analysis*. New York: Center for Chemical Process Safety of the American Institute of Chemical Engineers.

Andrews, J.D. (1993). *Reliability and Risk Assessment*. Harlow, Essex: Longman.

Boff, K.R. and Lincoln, J.E. (1988). *Engineering Data Compendium Human Perception and Performance*. Harry G. Armstrong Aerospace Medical Research Laboratory, Human Engineering Division, Wright Patterson Air Force Base, OH 45433, USA.

Booth, R.T. (1993). *Where's The Harm In It?* Risk Assessment Work Book. Henley-on-Thames: Monitor Training.

Central Statistical Office (1988). *Annual Abstract of Statistics, 124*. London: HMSO.

Cheyne, A.J., Cox, S.J. and Raw, G. (1994). 'Development of a risk assessment model for building use: scales for measuring severity of harm'. In *Conference Proceedings, Healthy Buildings '94, Budapest, August* (L. Bászló, I. Farkas, Z. Magyar and P. Rudnai, eds) pp. 89–94, Technical University of Budapest.

Covello, V.T. (1983). 'The perception of technological risk: a literature review'. *Technological Forecasting and Social Change*, **23**, 285.

Covello, V.T. (1989). 'Informing people about risks from chemicals, radiation and other toxic substances: a review of obstacles to public understanding and effective risk communication'. In *Prospects of Problems in Risk Communication* (W. Leiss, ed), Canada: Institute for Risk Research, University of Waterloo Press.

Cox, S. (1992). *Risk Assessment Toolkit*. Loughborough: Loughborough University of Technology.

Cox, S.J. and O'Sullivan, E.F. (1995). *Building Regulation and Safety*. London: Construction Research Publications Ltd.

Cox, S. and Tait, N.R.S. (1991). *Reliability, Safety and Risk Management – An Integrated Approach*. London: Butterworth-Heinemann.

Cox, S. and Tait, N.R.S. (1993). 'From risk analysis to risk management – the developing role of the engineer'. In *Engineers and Risk Issues, Proceedings of the Symposium of The Safety and Reliability Society*, Altrincham, UK, October.

Cox, S., Janes, W., Walker, D. and Wenham, D. (1995). *Office Health and Safety Handbook*. London: Tolley.

Cox, T. (1985). 'Repetitive work: occupational stress and health'. In *Job Stress and Blue Collar Work*. (C.L. Cooper and M. Smith, eds), Chichester: Wiley and Sons.

Cox, T. (1993). *Stress Research and Stress Management: Putting Theory to Work*. Sudbury, Suffolk: HSE Books.

Cox, T. and Cox, S. (1993). *Psychosocial and Organisational Hazards: Monitoring and Control. European Series in Occupational Health no. 5*. Copenhagen, Denmark: World Health Organization (Regional Office for Europe).

Cox, T., Ferguson, E., and Farnsworth, W.F. (1993). 'Nurses' knowledge of HIV and AIDS and their perceptions of the associated risk of infection at work'. Paper to: *VI European Congress on Work and Organisational Psychology*, Alicante.

Cox, T.R., Griffiths, A.G. and Stokes, A. (1995). 'Work-related stress and the law: the current position'. *Journal of Employment Law and Practice*, **2**, 93–96.

Douglas, M. (1992). *Risk and Blame*. London: Routledge.

Douglas, M. and Wildavsky, A. (1982). *Risk and Culture: An Essay on the Selection of Technical and Environmental Dangers*. Berkeley: University of California Press.

Engineering Employers Federation (EEF) (1993). *Practical Risk Assessment: Guidance for SMEs*. London: Engineering Employers Federation.

Fido, A.T. and Wood, D.O. (1989). *Safety Management Systems*. Shaftsbury: Blackmore Press.

Glendon, A.I. and McKenna, E.F. (1995). *Human Safety and Risk Management*. London: Chapman and Hall.

Griffiths, A., Buckle, P., Mackay, C.J. and Cox, T. (eds) (1996). *Work-Related Upper Limb Disorders: A Multidisciplinary Approach*. London: Taylor and Francis.

Hale, A.R. (1986). 'Subjective risk'. In *Risks: Concepts and Measures* (W.T. Singleton and J.J. Harden, eds) Chichester: John Wiley.

Handy, J. and Ford, M. (eds) (1993). *Redgraves, Fife and Machin: Health and Safety* (2nd edn). London: Butterworths.

Harrison, E.F. (1987). *The Organizational Decision-making Process*, 3rd edn. Boston: Houghton Mifflin.

HAZDATA (1990). National Chemical Emergency Centre, UKAEA, Harwell, Oxford, OX11 0RA, UK.

Health and Safety Commission (1992). *Management of Health and Safety at Work*. London: HMSO.

Health and Safety Executive (1988). *The Tolerability of Risk from Nuclear Power Stations*. London: HMSO.

Health and Safety Executive (1989). *Quantified Risk Assessment: Its Input Into Decision Making*. London: HMSO.

Health and Safety Executive (1992). *Display Screen Equipment Work. Health and Safety (Display Screen Equipment) Regulations 1992. Guidance on Regulations*. London: HMSO.

Health and Safety Executive (1994). *Essentials of Health and Safety at Work*. London:

HMSO.

Holroyd, K.A. and Lazarus, R.S. (1982). 'Stress, coping and somatic adaptation'. In *Handbook of Stress* (L. Goldberger and S. Breznitz, eds), New York: Free Press.

Howarth, C.I. (1987). 'Perceived risk and behavioural feedback: strategies for reducing accidents and increasing efficiency'. *Work and Stress*, 1, 61–67.

Humphreys, P. (1988). *Human Reliability Assessors' Guide*. Culcheth, Warrington: UKAEA.

International Labour Office (ILO) (1986). *Psychosocial Factors at Work: Recognition and Control. Occupational Safety and Health Series no: 56*. Geneva: International Labour Office.

Kaplan, S. and Garrick, B.J. (1981). 'On the quantitative definition of risk'. *Risk Analysis*, 1, 11–27.

Kazer, B (1993). *Risk Assessment: A Practical Guide*. Leicester: Institution of Occupational Safety and Health.

Kletz, T.A. (1971). 'Hazard analysis – a quantitative approach to safety'. *Institution of Chemical Engineers Symposium Series*, 34, 75–81.

Levi, L. (1981). *Preventing Work Stress*. Reading, MA: Addison-Wesley.

Levi, L. (1984).' Stress in Industry: Causes, Effects and Prevention'. *Occupational Safetyand Health Series*, 51 (Geneva: International Labour Office).

Marshall, V.C. (1987). *Major Chemical Hazards*. Chichester: Ellis Harwood.

Office of Population Censuses and Statistics (OPCS) (1991). *Mortality Statistics: Accidents and Violence*. London: HMSO.

O'Riordan, T. (1985). 'Approaches to regulation'. In *Regulating Industrial Risks* (H. Otway and M. Peltu, eds) pp. 21–39, Oxford: Butterworth-Heinemann.

Petts, J. (1989). 'Planning and major hazard installations'. In *Safety Cases* (F. Lees and M.L. Ang, eds), London: Butterworths.

Reason, J.T. (1990) *Human Error*. Cambridge: Cambridge University Press.

Ridley, J. (1993). *Safety at Work* (3rd edn) London: Butterworth-Heinemann.

Roberts, K.H., Stout, S.K. and Halpern, J.J. (1994). 'Decision dynamics in two high reliability military organisations'. *Management Science*, 40(5), 614–624.

Royal Society (1983). *Risk Assessment: A Study Group Report*. London: Royal Society.

Slovic, P., Fischhoff, B. and Lichtenstein, S. (1981).' Facts and fears: understanding perceived risk'. In *Society Risk Assessment. How safe is safe enough?* (R.C. Schwing and W.A. Alberts, eds) pp. 181–197, New York: Plenum Press.

Soby, B.A. and Ball, D.J. (1991). *Consumer Safety and the Valuation of Life and Injury*. London: Consumer Safety Unit, Department of Trade and Industry.

Steel, C. (1990). 'Risk estimation'. *The Safety and Health Practitioner*, June, 20–24.

Stevens, G. (1992). 'Workplace injury: a view from HSE's trailer to the 1990 Labour Force Survey'. *Employment Gazette*, December 100(12).

Teeling Smith, G. (1985). *Health Measurement*. London: Office of Health Economics.

Thomson, J.R. (1987). *Engineering Safety Assessment, An Introduction*. Harlow: Longman Scientific and Technical.

Torrance, G.W., Boyle, M.H. and Horwood, S.P. (1982). 'Application of multi-attribute utility theory to measure social preferences for health states'. *Operations Research*, 30, 1043–1069.

van der Schaaf, T.W., Lucas, D.A. and Hale, A.R. (eds) (1991). *Near Miss Reporting as a Safety Tool*. London: Butterworth-Heinemann.

Walker, D. and Cox, S.J. (1995). 'Risk assessment: training the assessors'. *The Training Officer*, July/August, 179–181.

Warner, F. (1992). 'Introduction: terminology'. In *Risk: Analysis, Perception and Management. Report of the Royal Society Study Group*. London: Royal Society.

Wilson, D.C. and Rosenfeld, R.H. (1990). *Managing Organisations: Text, Readings and Cases*. London: McGraw-Hill.

Woodall, F.J. and Cox, S.J. (1993). 'Stumbling into a serious accident'. *Health and Safety Practitioner*, May, 42–46.

Yates, D.W. (1990). 'Scoring systems for trauma'. *British Journal of Medicine*, **301**, 1090–1094.

3
Safety science and safety management

Introduction

Safety has been defined as 'a state of freedom from unacceptable risk of personal harm' (Fido and Wood, 1989). The meaning of this definition can be easily and simply explicated by pointing up its two key phrases: unacceptable risk and personal harm.

Although Fido and Wood (1989) grounded their definition of safety in terms of individual wellbeing, in current practice the concept is being applied as much at organizational and societal levels as it is at the individual level. Despite this extension in practice, the present text accepts and uses a notion of safety grounded in a challenge to individual wellbeing. Within this framework a safe system of work (see Chapter 11), for example, is one which is free from unacceptable threat to individual wellbeing. Threats to the wellbeing of organizations and to that of society are, in their place, proper and important concerns but are not seen by the present authors as the ultimate concern of this text. The concern here is for the *individual*. If *en route* some interest and concern is shown for the wellbeing of their organizations or of wider society, then that concern is expressed in relation to the effects of organizational and societal wellbeing on that of the individual. To this extent this text expresses a tradition rooted in western liberal philosophy (concern and freedom for the individual). Not surprisingly, and for many different reasons, the text also expresses, both implicitly and explicitly, the belief that the study and management of safety is a science – safety science – and the subject of logical and empirical enquiry. What we know about safety and how we apply and evaluate the practical value of that knowledge is best developed through the scientific method. Effective safety management is, at one and the same time, a child of science and the servant of the individual.

By introducing the notion of 'unacceptable' risk into their definition, and by implication 'acceptable' risk, Fido and Wood (1989) open the door to a safety science which recognizes both the inevitability of risk and the freedom of choice which individuals, organizations and society must exercise in the management process. The meaning of choice adopted here is essentially Aristotelian and implies analytical and decision making processes and the adoption of an action arrived at by such processes. The management of safety must thus incorporate some form of risk assessment and this, in turn, must develop through risk analysis and evaluation, with subsequent decision making providing the bridge into the management of unacceptable risks (see Chapter 2).

Risk of personal harm has been modelled in many different ways over the last six or so decades; some of the various models which have been described in the scientific literature have been briefly discussed in the previous chapter (see Chapter 2). At this stage, the important point to draw from these various models is the distinction, which might be best presented as a continuum, between acute and chronic risks. The event which in some way triggers or otherwise expresses the risk as personal harm is usually referred to as an accident if it was in some sense 'unplanned' and 'unpredictable'. The notion of the 'accident' has thus become central to much of traditional thinking about safety, and much of that thinking has focused on the question of accident causation. A certain narrowness has been associated with this traditional line of enquiry, an element of which has been a focus on the *'post hoc'*, and often structural, analysis of accidents and the development of engineering models of safety 'states' built on the avoidance of accidents which have already occurred. Another feature has been the focus on the role of the individual in the accident process rather than the organization – a failure to appreciate the 'context' and the importance of the individual in accident prevention. Such models may not be best placed to prevent future accidents given that such events are, by definition, unpredictable. Whatever their strengths and weaknesses, models of accident causation have a place in safety science, and this chapter in the present text offers a brief review of some of the more influential as a necessary framework for understanding that science.

Basic causation models: sequence models

The earliest models of accident causation were simple in conception and economic in their postulations of the mechanisms involved. These were, in essence, a sequence of events, each resulting from some prior condition or event and resulting in some subsequent event from the initiation of the chain to its conclusion in the accident. Such events were tied together in a deterministic sequence of cause and effect.

Heinrich (1931) is credited with one of the first of such accident models. His 'domino' theory is based on the assumption that accident causation can be described as a chain or sequence of events leading to an accident and its consequences. The model is illustrated in Figure 3.1. The five factors or predisposing stages in the sequence of events were:

1 hereditary and social environmental factors, leading to
2 a fault of the person constituting the proximate reason for
3 either an unsafe act or unsafe condition (or both) which results in
4 the accident, which leads to
5 the injury.

Heinrich (1931) likened these five stages in the sequence to five dominoes standing on edge in a line next to each other (see Figure 3.1), hence the 'domino' theory. He further postulated that, if the first domino falls it

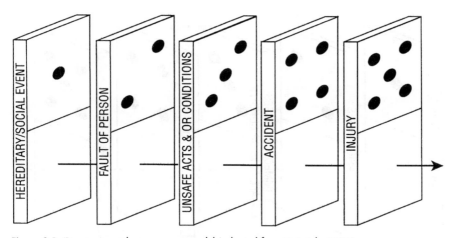

Figure 3.1 'Domino' accident sequence model (adapted from Heinrich, 1931)

automatically knocks down the second domino which in turn knocks down the third domino and so on. Removal of any one of the first four dominoes would break this sequence and so prevent personal injury. Using this metaphor, Heinrich (1931) argued that the key to prevention was the elimination of unsafe acts and/or unsafe conditions, the third domino in the sequence.

Interestingly, Heinrich's (1931) domino theory implies that 'the person' is largely responsible for 'carrying' the sequence of events to its conclusion in the accident: hereditary factors, a fault of the person, and an unsafe act. Such thinking undoubtedly supported the approach of early theorists

(Greenwood and Woods, 1919) which rather unhelpfully attributed 'blame' for accidents on individual workers, and spawned the notion of the accident prone worker (see later).

Several later authors have developed models based on or otherwise derived from Heinrich (1931). For example, Bird and Loftus (1976) developed Heinrich's model to reflect the influence of management in the cause and effect of all accidents that result in a wastage of the organization's assets. The outcome 'domino' is described more broadly by Bird and Loftus (1976) in terms of accident *consequences* – total loss. The most obvious losses are deemed to include harm to people, property or process (Bird and Germain, 1986). However, implied and related losses include 'business interruption' and 'profit reduction'. Bird and Germain (1986) further argue that once the sequence has occurred the exact nature and type of loss incurred is a matter of chance ranging from insignificant to catastrophic.

The modified sequence of events is summarized in Figure 3.2, the 'loss-causation' model.

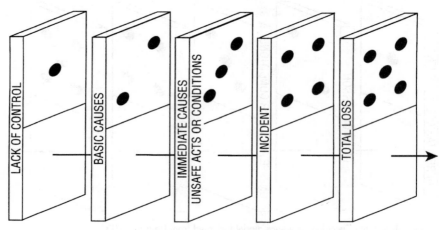

Figure 3.2 'Loss–causation' accident sequence model (adapted from Bird and Loftus, 1976)

The key feature, however, is their *first* 'domino' which focuses on *management control*. Bird and Germain (1986) describe control as one of the four essential management functions (planning, organizing, leadership and control). Lack of control is perceived to be a major factor in the accident or loss sequence and elimination of this 'defect' is seen to be central and the key to accident prevention. The introduction of the notion of 'control' into accident causation models is significant from, at least, three different points of view: engineering, management and psychological. The issue of control will be discussed in some detail later in this text (see, for example, the generic safety management model in Chapter 9).

Although both these models (Heinrich, 1931; Bird and Loftus, 1976) show causation as a single chain, their authors, at the same time, imply multicausality for accidents. However, the fact is that multicausality is not made explicit and that linear event sequences rather than systemic descriptions are stressed. This may introduce a bias and lead people to tendencies of fitting the available facts to a model in a restrictive manner. Rasmussen (1982) refers to this as the 'stop rule' which he describes as:

> an event will be accepted as a cause and the search (for further facts) terminated if the causal path can be followed no longer, or a familiar, abnormal event is found which is therefore accepted as an explanation, and a cure is known.

Expediency linked together with linear causation models have thus restricted many investigations. It has been further argued (Petersen, 1971) that by narrowing down the possible combinations of substandard acts and specific situations to the single stage, the identification and control of contributing causes has been severely limited. Recent researchers have therefore focused on multiple causation and multicausality.

Multiple causation theory (multicausality)

Multicausality refers to the fact that there may be more than one cause to any accident. Furthermore, it is often implied that such 'causes' summate in a logical fashion. Each cause may itself have multicauses and the process during accident investigation of following each back to its root is known as 'fault tree analysis' (see S Cox and Tait, 1991). This may be represented schematically using causal networks. The analogy of the tree is clearly reflected in the appearance of such networks (see Figure 3.5).

The 'nodes' in fault trees where causal lines join have been represented in 'summative' models as 'gates'. Two types of logic gates can be found in fault trees: AND-gates and OR-gates (see Figure 3.3).

Figures 3.4 and 3.5, adapted from S Cox and Tait (1991), illustrate the process diagram and the fault tree for the overturning of a container during movement by means of the lifting mechanism of another container. The conditions for a collision are the movement of a container unretracted *or* a collision with a triple stack. In the latter case the stack must have been incorrectly placed on the position *and* the operator must have failed to notice the error. For the container to be moved while unretracted, the retraction must be incomplete *and* movement must be attempted, while for movement to be attempted in these circumstances, there must be an operator error *and* a failure of the relevant interlock. The logic just described can easily be followed in Figure 3.5.

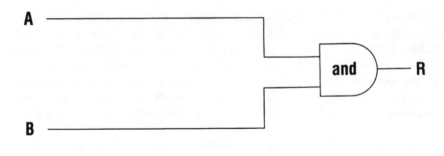

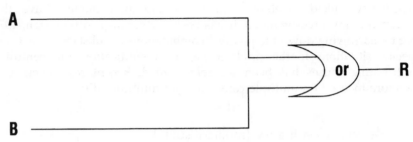

Figure 3.3 Representation of an AND-gate and an OR-gate

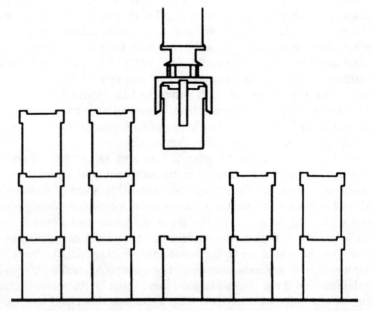

Figure 3.4 Container-handling system with three-high stacks (from S Cox and Tait, 1991)

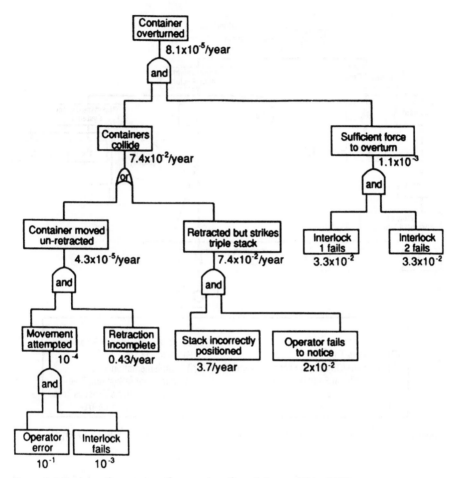

Figure 3.5 Fault tree for overturn of a container (from S Cox and Tait, 1991)

The logic is plotted in the fault tree (see Figure 3.5) and illustrates the fact that even 'simple' accidents have more than one cause.

In summary, the theory of multicausality is that the separate contributing causes combine together in a manner which results in an accident. It has been the basis of many separate models, including the Management Oversight and Risk Tree (MORT).

Management Oversight and Risk Tree (MORT)

The possible complexity of multicausal accident models is addressed by Johnson (1975) in MORT. The MORT fault tree (see Figure 3.6) is extremely detailed and has its origins in US aerospace and nuclear operations: the

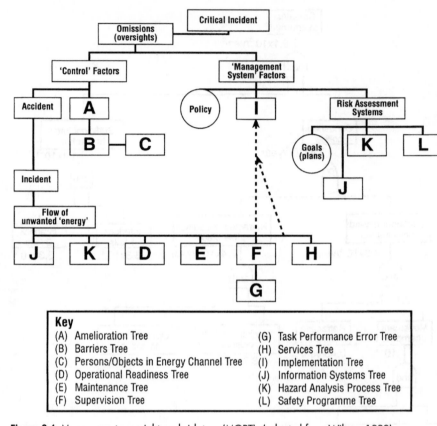

Figure 3.6 Management oversight and risk tree (MORT) (adapted from Wilson, 1993)

model identifies about 130 shortcomings in 'specific control' factors and over 100 'management system' factors. Wilson (1993) has described the application of MORT in the identification of human factors failures.

The complexity of the MORT logic tree reflects Johnson's (1973) theory of accident causation summarized as follows:

> Examination of occupational case histories suggests that antecedents often develop in a number of sequences involving physical, procedural and personal elements. Because the occupational setting is highly structured and controlled, we can look for the sequences of events that affected or changed the separate elements in every aspect of the industrial process, including:
> Management
> Planning and design
> Work environment (including arrangement and signals)

Machine (including tools, equipment and signals)
Materials
Supervision
Task procedure
Worker
Fellow worker (or other third party).
Frequently we find that a number of sequences were developing over a time before the interaction that produced the accident. Events in retrospect were on a collision course (Johnson, p.54, 1973).

Johnson's (1973) overview thus introduces the idea of system or process interactions into industrial accident causation and by implication 'dynamic' rather than static phenomena. It also introduces timescales which extend the period over which accident precursors developed and highlights the 'latency' factor as an important issue in safety as well as occupational health (World Health Organization (WHO), 1989). The advantage of such a model is that it focuses attention on deviations from 'normal system operating state' rather than single events. Such deviations are a consideration of the risk assessment system, which features (for the first time) explicitly in accident causation models.

'Normal states' and accident sequences

One of the current authors (S Cox and Tait, 1991) has adapted MacDonald's (1972) accident sequence model to incorporate the concepts of normal operation, deviation and recovery into the accident process (see Figure 3.7). She particularly stresses the importance of this model for the design of complex work technologies and the definition of 'normal systems' operation'. Potential 'accident' situations can be explored in the 'preplanning phase' using techniques such as failure modes and effects analysis (FMEA, described in S Cox and Tait, 1991) and recovery mechanisms (including operator intervention training) can be incorporated into the overall system at the design stage. This is usually the case in the design of complex technologies (see later) but is not always applied routinely in all workplaces where reactive rather than proactive approaches pertain.

The application of systems thinking to accident causation has been developed by many researchers and practitioners in this area, including one of the present authors (see, for example, S Cox and Tait, 1991; Hale and Glendon, 1987; Reason, 1995) and has assumed major importance in the safety arena. Systems theory is thus the focus of Chapter 4.

However, it is important here to stress the importance of the metastable state in the light of MacDonald's (1972) model (see Figure 3.7). This state, which follows a departure from 'normal' operation, may exist over quite a lengthy period. During the 'metastable' phase the system continues to

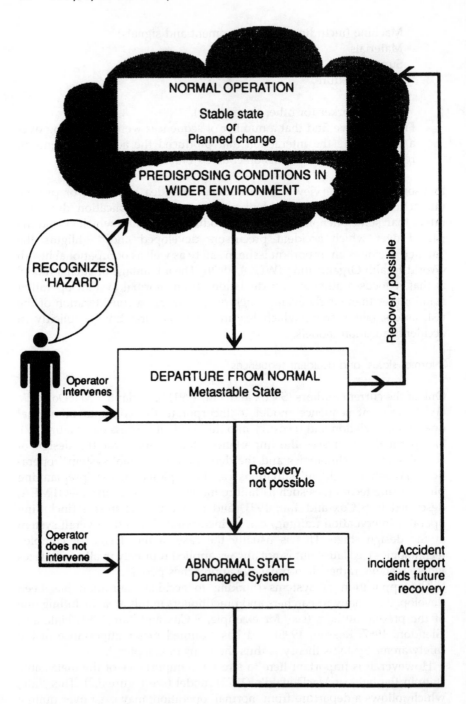

Figure 3.7 Systemic accident sequence model (from S Cox and Tait, 1991)

operate at an increased level of risk before any harm becomes imminent. Several researchers into accident causation (see, for example, Kjellen, 1984; Powell *et al.*, 1971; Raise, 1985; Smillie and Ayoub, 1976/7; Zohar and Fussfield, 1981) have stressed the role of complex organizational factors in suppressing detection of metastable states and have introduced the concept of 'systemic defences'. This has been further articulated in Reason's theory of latent and active failures (Reason, 1990a).

Latent and active failures

Research into recent organizational disasters has led Reason (1990a) to propose the theory of latent and active failures to distinguish two ways in which human beings contribute to their causation. He refers to *active* and *latent* failures in the following way:

Active failures
These are errors and violations (deliberate acts or behaviours) which have an immediate adverse effect on the system's integrity. They are committed by those directly involved in completing work tasks (commonly referred to as the 'sharp end' of production processes) and usually involve direct circumvention of system defences.

Latent failures
These are decisions or actions, whose damaging consequences may lie dormant for some considerable time, only becoming evident when they combine with local triggering factors (that is, active failures, technical faults, etc.) to break through the system's defences. Reason (1990b) has stressed their defining characteristic as their presence within the system well before the onset of a recognizable accident sequence. Furthermore, whereas active failures are more likely to be spawned by those at the 'sharp end', latent failures are likely to arise from these organizational activities which are removed in both time and space from the direct person-machine interface; for example, designers, high level policy and decision makers and managers.

Latent decisions and actions thus lie dormant in an organization for some time until triggered by active failures that are equivalent to the immediate substandard acts and conditions, shown in Figures 3.1 and 3.2. Reason further proposes a *resident pathogen* metaphor (Reason, 1988, 1990b); he postulates that latent failures are analogous to resident *pathogens* in the body which combine with external factors to bring about disease. Reason's theory may be summarized as follows:

1 'accidents' do not arise from single causes. They occur through the largely unforeseen concatenation of several distinct factors, each one necessary but singly insufficient to cause the catastrophic breakdown;

2 the likelihood of an accident is a function of the total number of pathogens (or latent failures) resident within the system;

3 the more complex, interactive, tightly coupled and opaque the system, the greater will be the number of 'resident *pathogens*';

4 simple, less defended, systems need fewer pathogens to bring about an accident;

5 the higher an individual's position within an organization, the greater is his or her opportunity for generating pathogens. The principal origin of resident pathogens are source types (see Figure 3.8) and they relate to the way in which senior managers chose to allocate finite resources between business or safety goals;

6 it is virtually impossible to foresee all the local triggers, though some could and should be anticipated. 'Resident *pathogens*', on the other hand, can be assessed, given adequate access and system knowledge; and

7 it therefore follows that the efforts of safety specialists could be directed more profitably towards the proactive identification and neutralization of latent failures, rather than at the prevention of active failures, as they have largely been in the past.

Reason's 'resident *pathogen*' metaphor has been incorporated into a model of accident causation (Reason, 1991) through the representation of organizational failure types (associated with classes of organizational and managerial failures) and individual failure tokens relating to failures at the person–machine interface (see Figure 3.8). Token failures are again synonymous

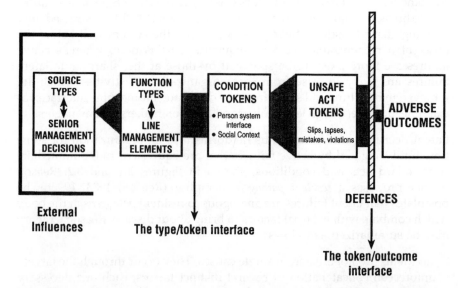

Figure 3.8 Reason's 'type–token–outcome' accident sequence model (adapted from Reason, 1991)

with the immediate causes, unsafe acts and conditions in the Bird and Loftus (1976) model (see Figure 3.2). However, Reason's model identifies two important interfaces, the type/token (or organizational/individual) interface and the token/incident (or individual/incident) interface (see Chapter 4). It also represents organizational failure types in terms of source types (Mintzberg, 1979) which are associated with fallible decisions at the highest level of organizations and functional types which relate to the line management element of failure.

Reason's (1991) model introduces the concept of 'amplification' in which source type failures (for example, failures to create a positive and optimistic safety culture) are communicated and distributed through the organization with a proliferation of incidents ((few/many) at the 'source/function' interface, and (many/many more) at the 'type-token' interface). This is represented in Figure 3.8 by increases in the shaded areas of the model. His model also emphasizes the complexity of interactions which means that the linking of specific adverse outcomes to 'source type' omissions or failures presents difficulties. Such associations or pathways can be considered through epidemiological methods. This approach links in with the WHO approach (1989) and is the basis of systems management approach.

The WHO safety management approach to accidents is that the immediate cause(s) of an accident (unsafe conditions and unsafe behaviour) are only symptoms of the true root cause(s) which exist in management function. For example, there may be errors in the areas of management policy, the setting of goals, staffing, housekeeping, responsibility, use of authority, line and staff relationships, accountability, and rules. The authors of the WHO document (1989) on the epidemiology of accidents point out that the process used in controlling the frequency and severity of accidents and controlling the quality and quantity of product have much in common (this is further explored in Chapter 9). In many cases, the same faulty practices are incriminated in both accident occurrence and unsatisfactory production. This link will be further explored in later sections of this book (see Chapter 9).

General failure types (linking failure mechanisms, substandard acts and accidents)

The earlier accident causation models (Heinrich, 1931; Bird and Loftus, 1976) did not focus on failure mechanisms but tended to focus on substandard acts or conditions. To a certain extent such models mirrored organizational methods of prevention which were more likely to be focused on 'operator–machine' interfaces than 'operator–management system' interfaces. However, in the light of recent disasters (for example, Chernobyl (1986), the capsize of the Herald of Free Enterprise (1987) and the Piper Alpha oil platform explosions (1988)) safety management has adopted a more systemic approach (as recommended, for example, by Cullen, 1990). This has been reflected in international research efforts (see, for example, the World Bank (1988) research programme workshops set up to determine critical management and organizational failures that may lead to catastrophic system failure). The proceedings of these workshops include the following introductory statement:

> The shift in research focus is to the operator/management system interface and to the critical internal/external management factors that can result in the propagation of a failure chain leading to a system failure.   (World Bank, 1988).

This relative importance of sources of failure against the general thrust of control strategies and the current approaches in major accident prevention is illustrated in Figure 3.9.

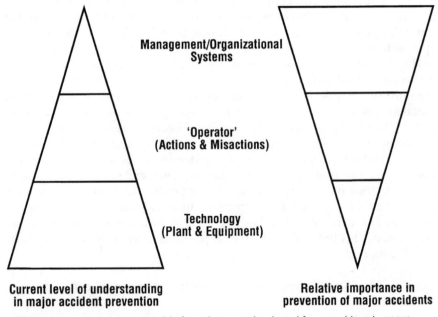

Figure 3.9 Triangular inversion model of accident control (adapted from World Bank, 1988)

This change in emphasis is reflected in Reason's (1991) model and in Tripod (Groeneweg, 1994). Tripod (Groeneweg, 1994) links accidents, unsafe acts and underlying causes through three interconnected nodes (see Figure 3.10).

Groeneweg's (1994) model represents the error generating mechanism and focuses attention on the presence or absence of general failure types. His contention is that it is unnecessary to identify all substandard (or unsafe)

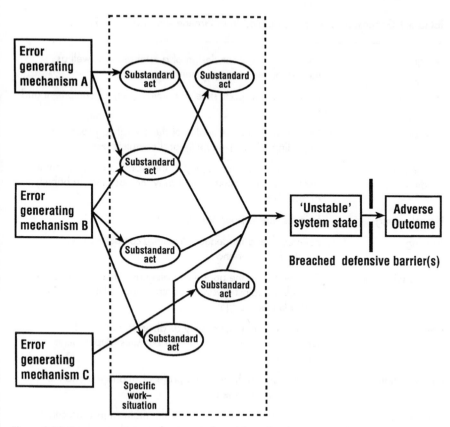

Figure 3.10 Error generating mechanisms (adapted from Groeneweg, 1994)

acts or clusters of substandard acts –the target for investigation is to discover whether any 'general failure types' are present (Groeneweg *et al.*, 1989). Indeed, the complexity of the interactions (or clustering) which is represented in Figure 3.10 renders such a task almost impossible. Table 3.1 illustrates the nature of general failure types used by Groeneweg and his colleagues (1989). When all the general failure types have been identified and error generating mechanisms have been plotted using Groeneweg's model, organizations should be able to ameliorate their impact (see later). This emphasis on 'sources' of failure rather than the accident itself has been further emphasized in some studies of pipework accidents commissioned by the UK HSE.

Sociotechnical pyramids

The sociotechnical pyramid (see Figure 3.11) developed by Technica (1989) during pipework accident studies and adapted by Harrison (1992) represents

Table 3.1 General failure types (adapted from Groeneweg *et al.*, 1989)

Design	Failures due to poor design of a total plant as well as individual items of equipment
Hardware	Failures due to poor state or unavailability of equipment and tools
Procedures	Failures due to poor quality of the operating procedures regarding utility, availability and completeness
Error enforcing conditions	Failures due to poor quality of the working environment in relation to circumstances that may increase the probability of mistakes
Housekeeping	Failures due to poor housekeeping
Training	Failures due to inadequate training/competence or insufficient experience
Incompatible goals	Failures due to the poor way safety and internal welfare are defended against a variety of other goals like time pressures and a limited budget
Communication	Failures due to poor quality or absence of lines of communication between the various sites, departments or employees
Organization	Failures due to the way the project is managed and the company is operated
Maintenance management	Failures due to poor quality of the maintenance procedures regarding quality, utility, availability and completeness
Defences	Failures due to the poor quality of the protection against hazardous situations

levels of increasingly 'remote' causes from the accident event – including, in this case (pipework integrity), an unplanned release, possible mitigation and eventual impact. This 'remoteness' is not necessarily temporal but relates to the number of intervening variables. The 'system climate' (at the base of the pyramid) represents the climate (both internal and external, see Chapter 4) in which the organization operates. The 'organization and management' (level four) refers to organization and management structures and systems. 'Communication, Information and Feedback Control' (level three) is focused on the impact of information dissemination and communication on systemic deviation. The information which flows from level three affects operator reliability, since it enables them to build up their own model of the system. Finally, level one refers to the engineered system and incorporates the

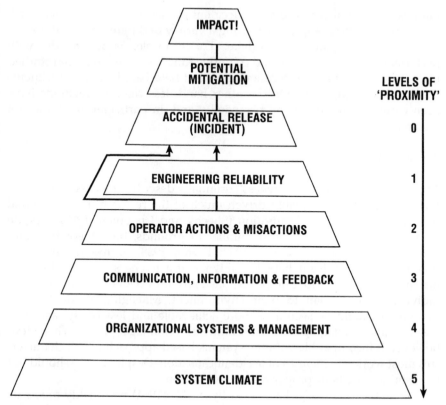

Figure 3.11 The sociotechnical pyramid (adapted from Harrison, 1992)

technology and systems of the plant and process. However, at this level the model only includes components of the system whose failure could lead directly to the accident (i.e. the unplanned release). Applications of the sociotechnical pyramid to the aetiology of major disasters are described in several technical reports on organizational, management and human factors (see, for example, Harrison, 1992).

Human behaviour and individual differences

Many of the factors described in models of accident causation explicitly involve, or at least imply the involvement of, aspects of human behaviour. In turn, such behaviour is often treated as the expression of cognitive and emotional functions or personality, and the discussion of their role is framed in terms of individual differences. Thus questions have been asked about the reliability of operator performance, human error, cognitive and attentional failures and distraction, stress and accident proneness. The answers to such questions have been approached using a variety of different methodologies, some more adequate than others. For example, issues to do with performance reliability, error and cognitive function have often been studied using laboratory-based experimentation and field-based quasi-experiments (Boff and Lincoln, 1988). On the other hand, less rigorous methods have been employed in studies on personality and, in particular, in studies on 'accident proneness'.

Accident proneness

The concept of 'accident proneness', originally described by Greenwood and Woods in 1919, has arguably driven much of the research into individual differences and accident causation (Sheehy and Chapman, 1987). Despite being severally criticized on a number of grounds, both theoretical and methodological (see McKenna, 1983; Porter, 1988), it has continued to dominate much of this debate. Possibly its attractiveness, in some quarters, lies in its implicit attribution of responsibility for accident causation to the individual rather than to their environment. Managing accident prone individuals might be seen as a more achievable and less costly task than creating and managing safe environments (Hollnagel, 1993). This view, however, is incompatible with the systems-based approach which characterizes most workable theories of accident causation. Despite this, it should not be dismissed without proper critical evaluation.

Shaw and Sichel (1971) have used Eysenck's (1964) personality model as a basis for relating personality characteristics to accident liability and found some relationship. For example, 'stable introverts' were deemed to be a 'good' risk in terms of their accident liability whereas 'unstable extroverts' were deemed to be a 'bad' risk and 'unstable introverts' were a 'poor' risk. The personality characteristics associated with 'unstable extroverts' include aggression, hostility and restlessness. These characteristics have been found in accident repeating individuals in several studies (see, for example, Smith and Kirkham, 1981).

McKenna (1983) has referred to two particular problems. First, much of the research into accident proneness has used inappropriate statistical techniques. As Boyle (1980) has pointed out, the unequal liability to accidents within the population is described reasonably well by a negative binomial distribution, and better than by a Poisson distribution. However, acceptance of a negative binomial distribution rests on the possibly untenable assumption that exposure and biases in accident reporting are equally distributed across the population in question (Sheehy and Chapman, 1987). Porter (1988) has also pointed to the difficulties of using a statistical approach to study human characteristics which may 'cause' accidents. Second, McKenna (1983) argues that the concept is used loosely and inconsistently in the literature. Attempting to reduce this inconsistency in and looseness of definition, while excluding methodologically unsound studies, dramatically reduces the evidence in favour of the concept of 'accident proneness' and leaves it as a hypothesis still to be tested rather than a fact and an explanation for accident events.

McKenna (1983) suggests that a more neutral term, differential accident involvement, should be used instead of accident proneness as it may be more neutral and, in being so, weaken the practice of attempting to identify individuals or small groups *responsible for* accidents. This suggestion has been supported by Sheehy and Chapman (1987).

Differential accident involvement

If individuals are not more or less accident prone, in terms of their personality or habitual ways of behaving, then what role does the individual and their behaviour play in the accident sequence? Several different roles have already been implied by the theories reviewed above. One relatively common approach is that the individual's behaviour is shaped by their environment, but that cognitive and emotional mechanisms drive that process (see Chapter 7). Therefore, individual involvement in the accident sequence cannot be treated out of the context of their environment and their interactions with that environment. We might therefore phrase questions about individual differences and behaviour in terms of the nature of the cognitions which determine the person's behaviour within its environmental context, and of the emotions which reflect and colour those cognitions and which influence behaviour. This type of approach, characterized by environmental stimuli, cognitive function and performance (behaviour), is represented, for example, in the work of Reason (1990a) and Groeneweg (1994) (described earlier).

Adopting this framework, it becomes clear that the individual is, in part, a carrier of the effect of the environment in relation to accident causation. The assumption is that, by and large, a safe environment determines a low accident rate while, conversely, an unsafe environment determines a high accident rate. Errors are largely a reflection of an unsafe environment and the individual making those errors is largely in the wrong place at the wrong time. Removing the individual who makes errors from an unsafe environment will not substantially change the probability of an accident as others will be or will become carriers of the effect of that environment. Managing this scenario effectively is a matter of improving the design of the environment – removing or reducing the conditions which make it an unsafe system – removing, in Groeneweg's (1994) terms, the 'error generating mechanisms' in the system.

While it is appealing to those who wish to absolve the individual of all responsibility for accident causation, this approach is probably too extreme a reaction to the notion of accident proneness. What is undoubtedly more sensible is a compromise model which recognizes, within the approach just described, some room for the operation of individual differences (see Chapter 7). This position may be made clearer by developing the medical analogy of the individual as a carrier of the effects of the environment in the accident sequence.

The carrier analogy

An unhealthy environment breeds disease, or at least supports the development of disease agents. These are transmitted within and between species by carriers. Destroying existing individual carriers will not stop the transmission of the disease as other individuals may become infected and act as carriers. Only by dealing with the source of the problem – the unhealthy environment – will the disease be eradicated. This is essentially the argument made out for the role of the individual in the accident sequence.

However, developing the carrier analogy, not all individuals exposed to the unhealthy environment become carriers. Some are more resistant than others and, in turn, resistance is determined by a multitude of factors acting together. The question is what are the factors which sensitize the individual to the effects of their environment or increase their resistance to it? Obviously, the bald concept of accident proneness has not yet been proved to be one of them. Indeed, asking questions about personality factors is probably not the most productive way forward.

Recent research in this area has adopted a new methodological approach based on the use of advanced multivariate techniques and, in particular, path analysis (Kenny, 1979) and structural equation modelling such as LISREL (Jöreskog and Sörbm, 1984). These techniques allow correlational data to be modelled in a way which implies directionality in relationships and which allows for the testing out of the adequacy of different hypothesized models. An example of this approach applied to the relationships between accidents and individual differences in biodata, personality and cognitive factors has been reported by Hansen (1989).

Hansen (1989) attempted to construct and test out a causal model of the accident sequence. Data were gathered from 362 chemical industry workers using questionnaires. The suggested causal model was analysed and cross-validated using LISREL-VI. It was proposed that:

1 employee age and job experience;
2 level of social maladjustment and distractibility; and
3 cognitive ability

would have independent causal effects on the accident criterion, even when differential accident risk and involvement in counselling were controlled for.

The results showed that the causal model as a whole was viable in the initial and cross-validation analyses, and the social maladjustment and distractibility variables were shown to be important causal factors for accidents. The role of cognitive ability was less clear and was confounded in the study by its earlier role as an important selection criterion. Use in selection may have resulted in a restriction of range in the cognitive ability scores of the population studied, and restriction of range effects in the LISREL analysis which removed any correlation between cognitive ability and accidents. Hansen (1989) noted that his results could not be generalized with any certainty and may be situation specific. In different situations, other psychological or behavioural variables might demonstrate importance as causal factors for accidents. What is important, that author claimed, is the methodology used. This raises questions about the utility of accident causation models not derived using this methodology or similarly adequate methodologies. Their role in safety management systems should be critically evaluated.

Accident causation models: their role in safety management

The early application of accident causation models to safety management was based on an approach which 'reactively' managed accidents and incidents rather than their underlying causes. The key features of this approach were:

1 the search for a single (primary) accident cause; and
2 the debate as to whether the primary cause was an unsafe act or an unsafe condition.

A simplified view of this reactive process is presented in Figure 3.12 (adapted from Booth, 1995). Figure 3.12 illustrates how management actions for safety traditionally involved either the preparation of a safety rule designed to prevent a recurrence of the 'unsafe act', or a physical safeguard to remedy the 'unsafe condition' most proximate to the accident. It always raised the question as to the relative contribution of each of these causes to the accident; which branch should the accident investigator or manager follow through?

The relative contribution of unsafe acts and unsafe conditions in accident causation has been the topic of prolonged debate. The emphasis given to unsafe acts culminated in the work on accident proneness carried out by Hale on behalf of the Robens Committee (Hale and Hale, 1970). Supporters of this approach argued that accident proneness is a detectable and predictable personality trait, and that the prevention can be achieved by selecting out the accident prone. However, research into this concept (Porter, 1988) has revealed problems with the earlier studies including statistical

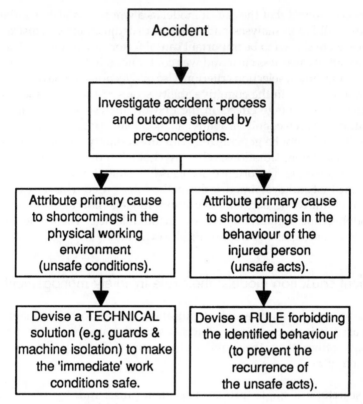

Figure 3.12 Application of 'traditional' models of accident causation to safety management (adapted from Booth, 1995)

weaknesses in analyzing the data. It has also been argued more generally (ACSNI, 1993) that the 'causation-debate' has often missed three critical and interrelated issues that:

1 the concept of a single primary accident cause is a simplification of a complex multi-causal process;
2 the focus on causality can mask the importance of worker behaviour in support of accident prevention; and
3 the arguments have focused almost exclusively on the errors made by the workers who have had the accidents rather than the errors made by managers and engineers whose behaviour (remote in time and space from the accident location) may have contributed to the prevailing safety climate (or culture).

More adequate accident-causation models, including MORT (Johnson, 1975), theories of latent and active failures (Reason, 1990a, 1990b) and error generating mechanisms (Groeneweg, 1994) and, finally, sociotechnical and systems theory (S Cox and Tait, 1991; Technica, 1989) have developed the above issues and can provide more effective frameworks for understanding accidents. At the same time, developments in statistical analysis have allowed a more thorough exploration of the available data.

Summary

This chapter has attempted a review of models of accident causation as a framework for understanding safety science. Although the authors accept that such models may be not best placed to prevent future accidents given that such events are, by definition, unpredictable, they argue for an approach which exploits their utility in safety management.

Furthermore, the management of safety in the 1990s demands a more enlightened approach which works beyond the immediate causes (unsafe acts and conditions) and focuses on the total work system. Management actions in support of safe systems should be proactive and based on sound decision making and risk assessment. Senior managers and policy makers should recognize the importance of individuals (at all levels) in creating a safe organization and seek to develop systems which support (rather than ameliorate) safe working.

The next chapter describes a systems framework for the management of safety and provides a rationale for the authors' approach to this important aspect of organizational control.

References

Advisory Committee for Safety in Nuclear Installations (ACSNI) (1993). *Third Report, Organising for Safety, of the Human Factors Study Group of ACSNI*. Sudbury, Suffolk: HSE Books.

Bird, F.E. and Germain, G.L. (1986). *Practical Loss Control Leadership*. Loganville, GA: Institute Press.

Bird, F.E. and Loftus, R.G. (1976). *Loss Control Management*. Loganville, GA: Institute Press.

Boff, K.R. and Lincoln, J.E. (1988). *Engineering Data Compendium, Human Perception and Performance*. Harry G. Armstrong Aerospace Medical Research Laboratory, Human Engineering Division, Wright-Patterson Air Force Base, OH 45433, USA.

Booth, R. (1995). The role of human factors in safety management. Paper presented to I.I.R. Conference, *The Practicalities of Developing and Mastering Proactive Safety Management*, January, London.

Boyle, A.J. (1980). 'A model of accident liability based on the normal distribution'. Paper presented at *The British Psychological Society's London Conference*, December.

Cox, S.J. and Tait, N.R.S. (1991). *Reliability, Safety and Risk Management*. Oxford: Butterworth-Heinemann.

Cullen, D.W. (1990). *The Public Inquiry into the Piper Alpha Disaster*. London: HMSO.

Eysenck, H.J. (ed) (1964). *Experiments in Motivation*. Oxford: Pergamon.

Fido, A.T. and Wood, D.O. (1989). *Safety Management Systems*. Shaftsbury: Blackmore Press.

Greenwood, M. and Woods, M. (1919). *The incidence of industrial accidents upon individuals with special reference to multiple accidents*. Industrial Fatigue Research Board (Report No. 4., 1919). London: HMSO.

Groeneweg, J. (1994). *Controlling the Controllable*. The Netherlands: DSWO Press, Leiden University.

Groeneweg, J., Hudson, P.T.W. and Waagenar, W.A. (1989). *Towards a complete list of general failure types*. Interim report prepared for Shell Internationale Petroleum Maatschappi, Exploration and Production.

Hale, A.R. and Glendon, A.I. (1987). *Individual Behaviour in the Control of Danger*. *Industrial Safety Series*, Vol 2, Holland: Elsevier.

Hale, A.R. and Hale, M. (1970). 'Accidents in perspective'. *Journal of Occupational Psychology*, **44**, 115–121.

Hansen, C.P. (1989). 'A causal model of the relationship among accidents, biodata personality and cognitive factors'. *Journal of Applied Psychology*, **74**(1), 81–90.

Harrison, P.I. (1992). *Organisational, Management and Human Factors in Qualified Risk Assessment, Report 2*. (HSE Contract Research Report No. 34, 1992). London: HMSO.

Heinrich, H.W. (1931). *Industrial Accident Prevention* (4th edn, with D. Petersen & N. Ross. McGraw Hill, New York, 1969).

Hollnagel, E. (1993). *Human Reliability Analysis: Context and Control*. London: Academic Press.

Johnson, W.G. (1973). 'Sequences in Accident Causation'. *Journal of Safety Research*, **5**(2), 54.

Johnson, W.G. (1975). 'MORT: The Management Oversight and Risk Tree'. *Journal of Safety Research*, **7**(1), 5–15.

Jöreskog, K.G. and Sörbm, D. (1984). *LISREL VI*. Chicago: International Educational Services.

Kenny, D.A. (1979). *Correlation and Causality*. New York: Wiley.

Kjellén, U. (1984). *The deviation concept in occupational accident control – 1. Accident Analysis and Prevention*. TRITA/AUGD019 Arbetsolyckfalls Gruppen. Stockholm: Royal Institute of Technology.

MacDonald, G.L. (1972). *The involvement of tractor design in accidents*. (Research Report 3/72, 1972). St. Lucia: Department of Mechanical Engineering, University of Queensland.

McKenna, F.P. (1983). 'Accident proneness: A conceptual analysis'. *Accident Analysis and Prevention*, **15**, 65–71.

Mintzberg, H. (1979). *The Structuring of Organisations*. Englewood Cliffs, NJ: Prentice-Hall.

Petersen, D. (1971). *Techniques of Safety Management*. New York, McGraw-Hill.

Porter, C.S. (1988). 'Accident proneness: a review of the concept'. In *International Reviews of Ergonomics: Current Trends in Human Factors Research and Practice, Volume*

2 (D.J. Oborne, ed) pp. 177–206, London: Taylor and Francis.

Powell, P.I., Hale, M., Marks, J. and Simon, M. (1971). *2000 Accidents: A Shop Floor Study of Their Causes*. London: National Institute for Industrial Psychology.

Raise, N.B. (1985). 'Models of human problem solving. Detection, changes and compensation for system failures'. *Automatica*, **19**, 613–625.

Rasmussen, J. (1982). 'Human errors. A taxonomy for describing human malfunction in industrial installations'. *Journal of Occupational Accidents*, **4**, 311–333.

Reason, J.T. (1988). 'Modelling the basic error tendencies of human operators'. *Reliability Engineering and System Safety*, **22**, 137–153.

Reason, J.T. (1990a). *Human Error*. Cambridge: Cambridge University Press.

Reason, J.T. (1990b). 'The contribution of human failures in the breakdown of complex systems'. *Philosophical Transactions of the Royal Society* (B), **327**, 475–484, London.

Reason, J.T. (1991). 'Too little and too late: a commentary on accident and incident reporting systems'. In *Near Miss Reporting as a Safety Tool* (T.W. van der Schaaf, D.A. Lucas and A.R. Hale, eds) London: Butterworth-Heinemann.

Reason, J.T. (1995). 'A systems approach to organizational error'. *Ergonomics*, **38**(8), 1708–1721.

Shaw, L.S. and Sichel, H.S. (1971). *Accident Proneness*. Oxford: Pergamon.

Sheehy, N.P. and Chapman, A.J. (1987). 'Industrial accidents'. In *International Review of Industrial Psychology* (C.L. Cooper and I.T. Robinson, eds) pp. 201–227, J. Wiley & Sons Ltd.

Smillie, R. and Ayoub, M. (1976/7). 'Accident causation theories: a simulation approach'. *Journal of Occupational Accidents*, **1**, 47–68.

Smith, D.I. and Kirkham, R.W. (1981). 'Relationship between some personality characteristics and driving record'. *British Journal of Social Psychology*, **20**, 299–331.

Technica (1989). *Evaluation of the Human Contribution to Pipework and In-Line Equipment Failure Frequencies*. Final Report to HSE. London: Technica.

Wilson, H.C. (1993). 'The use of the Management Oversight Risk Tree in the identification of human factor failures'. In *Proceedings of the Third Annual Conference on Safety and Well-Being at Work* (S. Cox and A. Cheyne, eds) pp. 115–127, Loughborough: Centre for Hazard and Risk Management, Loughborough University of Technology.

World Health Organization (WHO) (1989). *Epidemiology of work related diseases and accidents*. Geneva: WHO.

World Bank (1988). *Workshop to develop a multi-sectoral/multi-disciplinary research program to determine critical management and organisational failures that may lead to catastrophic system failure*. World Bank, 1818 H Street, N.W. Washington DC 20433, USA.

Zohar, D. and Fussfield, N. (1981). 'A systems approach to organisational behaviour modification: theoretical considerations and empirical evidence'. *International Review of Applied Psychology*, **30**, 491–505.

Part 2
Psychology of Safety

The second part of the book considers the psychology of safety, and does so from a systems theory perspective. Systems theory is used to provide both the language and a framework for an exploration of the person in their job in the organization.

Organizations are discussed as systems and, in particular, as socio-technical systems. This approach is enriched by the introduction of new developments in systems thinking, complexity theory and the theory of organizational health, and of new, conceptual, tools such as the use of metaphors. The job is discussed in terms of its characteristics and constituent tasks and, the person is explored as an active information processor. Finally, the question of work-related stress is addressed as an indicator of systems failure.

The five chapters in Part 2 are:

4 A systems framework for the management of safety;
5 The organization;
6 Jobs and tasks;
7 The person;
8 Work-related stress.

4

A systems framework for the management of safety

Introduction

The effective management of any aspect of an organization requires forward planning, involving both vision and strategy, as well as the capacity to respond to events as they occur. Vision and strategy are related and together define the mission of the organization – what it wishes to achieve and how. All three are higher order aspects of the organization's function, and also determine its operational characteristics – the systems and tactics that it requires both to shape and to respond to events. Most texts on organizations deal with these variables and their interrelationships. However, what some of them miss is the fact that in order for an organization to develop a vision and a strategy for managing any particular function, it must have a way of thinking about it. The organization must have a conceptual framework for managing that function. This chapter is concerned with the conceptual framework adopted in this text for the management of safety at work.

Conceptual framework for managing safety

In their preface, and in the early chapters of this text, the authors have argued for a conceptual framework which combines three different perspectives:

1 an over-arching systems theory perspective which focuses, in part, on
2 the person in their job in their organization and which
3 treats the person not as a passive, but as an active, processor of hazard-related information conditioned by a plethora of psychosocial factors.

The title of this book – *Safety, Systems and People* – has been chosen to capture this interplay of ideas and, at the same time, to mark the continuation of the thesis established by Sue Cox's earlier book with Robin Tait: *Reliability, Safety and Risk Management* (1991). The present book develops a systems-based view of the interaction between a wide range of psychological, social and organizational factors underpinning the effective management of safety at work. The essential features of this model are presented in Figure 4.1. This

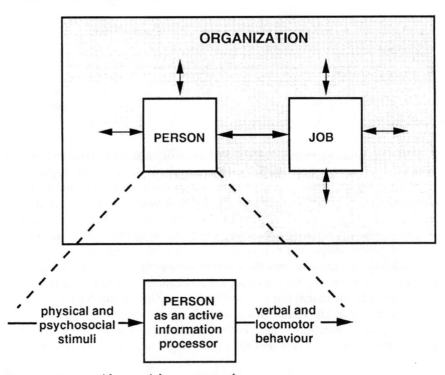

Figure 4.1 Conceptual framework for managing safety

conceptual framework for the management of safety is consistent with both systems theory approaches to safety management (for example, Waring, 1989) and the UK Health and Safety Executive's (HSE's) currently espoused approach, which is described below.

In 1989, the HSE published initial guidance on *Human Factors in Industrial Safety*. In the HSE's view (1989), human factors considerations encompass a wide range of individual, social and organizational issues. These include not only issues reflecting the individual's physical and psychological capabilities and constitution, but also those reflecting their interactions with the design and management of work systems, work technologies and work environments, and the characteristics of organizations as they relate to safety. The framework adopted by the HSE for making sense of such a diversity of issues is 'the individual in their job in their organization'.[1]

Two examples of the necessary sense of the authors' framework are suggested in the work of S Cox and her colleagues (S Cox, 1990 and S Cox et al., 1995).

The first example relates to the design of management control systems for safety in the chemical industry (S Cox, 1990). The evidence suggests that to be effective, the design of such systems must take account of the nature of the jobs covered, the abilities and other characteristics of the staff involved, the nature of staff-management relations, and the very structure and culture of the organization. It has been argued that systems which are designed without due regard for all these factors may have in-built weaknesses, and thus the potential for failure. Reason (1990) has called this sort of system weakness a 'latent failure' and sees it as a major contributor to the complex aetiology of certain types of accident (see Chapter 3).

The second example relates to the introduction of new safety technology into the office environment (S Cox et al., 1995). The evidence suggests that, to be successful, the introduction of such technology must consider the needs and abilities of the users, the nature of their social group, including its willingness to accept change, their preparation for the new technology and change (particularly in relation to the provision of information and training), and the nature of the change management strategy adopted. This, in turn, must reflect several of the other factors cited above, including the nature of the work to be done, how it will be changed by the new technology, and the culture of the organization. Failure to plan the introduction of any new technology taking these various factors into account will inevitably presage failure.

In both examples, it is not only attention to the individual factors which is required, but also attention to the way they *interact*. Thus, in the first example, it is not just the nature of the jobs covered which is important but also the extent to which the operators' abilities are *matched* to the demands of the job, and the extent to which those operators have control over their work and support in that work (see T Cox and Griffiths, 1995: Chapter 8). Furthermore, the operators' willingness to make that 'match' work will be conditioned by the nature of staff–management relations. In a 'healthy' organization, a poor fit might be made to be relatively effective through increased effort, while in an 'unhealthy' organization there might be a

---

[1] Arguably, the 'Human Factors' document (HSE, 1989) is the original source of the phrase 'the individual in their job in their organization', although similar phrases are to be found earlier in the literature.

refusal to exploit a good match. The concept of organizational health is dealt with later in this chapter, and in Chapter 5. Overall, staff–management relations may moderate the effectiveness of the operator–job match, while the importance of job characteristics can only be understood when related to those of the operators.

The principle of 'interaction' is also important in the second example and throughout the rest of this text.

An integrated approach to safety management

Among other things, the *Human Factors in Industrial Safety* document (HSE, 1989) appears to promote an *integrated* approach to safety management. Two levels of possible integration can be identified.

First, there is an argument for integration in the management of safety at a *higher level* by bringing together engineering systems and controls for plant and equipment (hardware), not only with effective management systems and procedures (software), but also with an understanding of and ability to manage people at work (see Figure 4.2).

The argument in favour of this level of integration is strengthened by the common observation that the majority of accidents are not solely attributable to 'human error' but to a complex interplay of human systems and technological failure (see Chapter 3). This important point has been reinforced in several HSE publications (see, for example, HSE, 1985, 1987). In 1987, the HSE published a study of maintenance accidents in the chemical industry and its recommendations for their prevention. It is clear from this publication that accident aetiology is largely an interplay of factors and that operators are not solely at fault. Indeed, it is recognized that the proposition that operators could be solely at fault is most unlikely given the complex and largely automated nature of the plant and technology used in the chemical industry. Despite the obviousness of this point, when stated, most organizations when they are attempting to improve safety performance address them as if the factors involved were separate and organized hierarchically. Initial attempts to improve safety tend to focus on aspects of hardware followed by consideration of the software (systems) involved. Usually, it is only when no further improvement can be made using this strategy, or when none has been apparent, that human factors are addressed (see Figure 4.3).

Second, there is an argument for integration in the management of safety at a *lower level* by identifying and bringing together the various and different human factors considerations through a single model. This, of course, can be achieved by using the framework of 'the individual in their job in their organization'. While the nature of the individual components of this framework is important, the most challenging issue intellectually, is understanding the activities which characterize the interface between any two or more components.

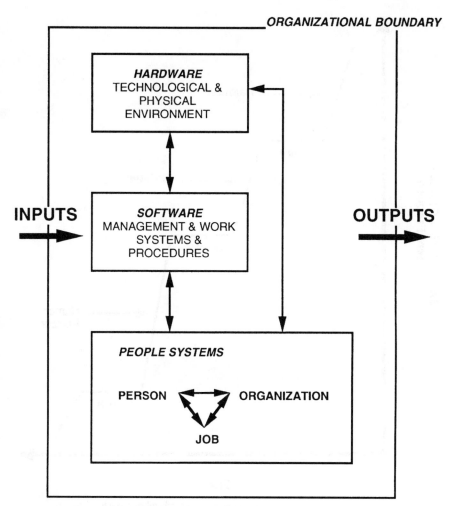

Figure 4.2 Safety systems: an integrated approach

Interfaces

There are three obvious two-way interfaces which help define this lower level of integration, those between: the organization and the person, the organization and the job, and the person and the job. Each of these interfaces can be considered from a number of different perspectives including those defined by the other factors involved at the higher level of integration: the technology (hardware) and the management systems (software). The overall framework which emerges is not dissimilar to that proposed in the relatively early work of Leavitt (1965): organization–jobs(tasks)–people–technology.

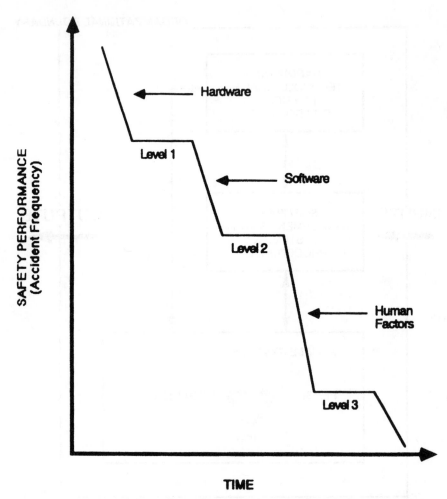

Figure 4.3 Improving workplace safety performance: a hierarchical approach

All this adds up to a complex system. Taken separately each component in this system may behave in compliance with relatively simple laws but when taken in combination, as a whole system, their behaviour may be less predictable and less controllable. It is the unpredictability and uncontrollability of the behaviour in organizations which challenges the psychology and management of safety.

The remainder of this chapter considers the role played by systems theory in the psychology and management of safety, and in particular in relation to our understanding of organizations and the role they play in determining safety at work (see Chapter 5). Subsequent chapters consider the organization (5), the job (tasks) (6), and the person (7). The chapter discussing the person treats individuals as active information processors. Together these chapters explicate the framework described above. This framework should allow us to describe, explain and predict the behaviours of complex organizational systems and, thus, provide a conceptual tool for the management of safety. Systems theory provides the 'way of thinking' which is required to make sense of and exploit the authors' framework for the psychology and management of safety.

Systems

A system is defined in common usage – see the *Oxford English Dictionary* (Little *et al.*, 1992) for example – as 'a whole composed of parts in an orderly arrangement according to some scheme or plan'. While this definition captures the structural aspects of systems, it perhaps lacks the 'feel' of the associated functional aspects. By and large, formal arrangement or organization supports activity. Systems do things, and one can talk meaningfully about the behaviour of systems. Indeed, schemes and plans are formulated in terms of objectives to be achieved through action. Formal arrangement (or organization) exists to make possible and support such action. Organization and behaviour thus represent the structure and function of systems.

Systems are complex, to a greater or lesser degree, and the orderly arrangement of systems components referred to by the *Oxford English Dictionary* may also exist to a greater or lesser extent and in a variety of forms. The question of complexity is discussed later in this chapter.

In relation to safety, systems can be defined as 'interacting sets of components forming hierarchies and networks for the purpose of fulfilling safety related objectives'. Such systems will be part of the total work organization. Treating the total work organization as *the* system, the safety system referred to here would have the status of a sub-system. The safety sub-system, and the various other sub-systems, would be arranged in some orderly way according to a structural and/or functional 'blueprint' to form the total work organization. In turn, each sub-system, safety-related or otherwise, might be broken down and analysed in terms of a number of further sub-systems in orderly arrangement according to some plan. The components of each system or sub-system may, in turn, be treated as systems in their own right, and so on. Important questions arise about limits to the deconstruction of any system of interest, about the boundaries of such a system and its components and function, and the wider environment in which it exists.

The basic concepts used to describe a generalized system are illustrated in Figure 4.4. The terms used in this figure are system elements (otherwise: components), relationship, boundary, input and output, wider environment

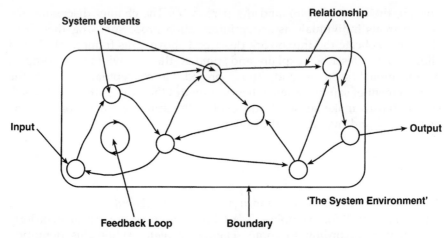

Figure 4.4 Generalized system diagram (adapted from Flood and Jackson, 1991)

and feedback. However, there are still other notions which are essential to a complete conception of a system. These include control, transformation, open system, homeostasis, emergence, communication, identity and hierarchy (see, for example, Carter *et al.*, 1984; Checkland, 1981; Flood and Jackson, 1991).

Components and boundaries

All systems are defined in terms of arrangements of boundaries and linked elements or components.[2] These are their relatively fixed parts. Components are related through 'connecting links' which indicate flows, influences and causal connections. These connecting links bring the system's structure to life and represent its 'process'. The system's boundary marks off these components and related processes from their wider environment and from those things which are unrelated and irrelevant to the system (see Figure 4.4).

The concept of a boundary is most obvious in relation to biological or engineering systems where it may have some physical basis. It is less obvious when discussing work organizations when it may be more conceptual than physical. The essence of a boundary is that it is the interface between different states (Flood and Jackson, 1991). The surface of a puddle is the boundary between the water forming that puddle and the air above it.

---

[2] Systems are collections of elements which may be grouped into sub-systems to give a picture of their structure. 'Component' is a general collective term for sub-systems and elements.

The border between England and Wales is the boundary which separates two of the geopolitical entities which comprise the British Isles. The former boundary has particular physical properties; the latter has not but has, instead, particular political meaning. Boundaries therefore assume particular properties or meanings. These properties or meanings determine the permeability of the boundary and nature of the transport of things across it. The characteristics of boundaries may change and the boundary may assume an enhanced form as when a wall is built around a work site or along an international border, and this may, in turn, affect its permeability. Changes occur in the characteristics and permeability of boundaries. This is as true for the passage of sodium ions across the axonal membrane as it was for the passage of citizens of the former Soviet Union and satellite states across the 'Iron Curtain'.

According to Carter *et al.* (1984) there are three principles which affect the definition of systems boundaries. First, a boundary excludes all components or relationships that have no functional effect on the system. Second, it includes all components or relationships which are strongly influenced or controlled by the system. Those that influence the system but are not influenced or controlled by it are part of its wider environment. Those that are irrelevant in all respects are outside the system model. Third, boundaries are defined in a way which either includes or excludes complete components rather than cutting across them. This minimizes the number and complexity of cross boundary relationships. Carter *et al.* (1984) also discuss the difference between closed and open systems, which is an issue of the obviousness and strength (permeability) of their boundaries.

Closed and open systems

A totally closed system is completely self contained and has no wider environment at all. It is not influenced by external events, nor does it, in turn, influence them. It serves no useful external purpose. Where growth is possible, a closed system has an inherent tendency to grow towards maximum homogeneity of its parts and it has been argued that a steady state may only be achieved by the cessation of all activities (Emery and Trist, 1981). In contrast, a totally open system is one in which the wider environment is so important that the system merges into it. Such a system has an arbitrary boundary and thus no stable identity; it is therefore difficult to plan for and manage. Open systems have to work to sustain their identity and maintain themselves in a steady state, often in a turbulent wider environment. Where growth is possible, open systems grow through a process of elaboration (Herbst, 1954) and may spontaneously reorganize towards states of greater complexity and heterogeneity (Emery and Trist, 1981). In reality, most work organizations are neither totally open nor totally closed but somewhere in between.

Internal processes

The system identified by a boundary will have inputs and outputs which may be physical or abstract, and that system works to transform those inputs into outputs. Inputs and outputs are transported across the boundary to and from the wider environment. The transformation process and the products of that process both represent and are represented in the purpose or objectives of the system. The three processes – input, transformation and output – require management and control, not only in terms of the various components and activities that they involve but also in terms of their relations with each other (see Figure 1.1 in Chapter 1). Internal processes are characterized by feedback and feedforward, whereby the behaviour of one element may influence, either directly or indirectly, that of another.

In summary, so far, systems, including organizations, can be described and understood in terms of their structures and functions. The description of structure can begin with the identification of the system's components (sub-systems and elements), the relationships between those components, its boundaries, and its inputs and outputs. The system's function is captured in its objectives, and the processes which support achievement of those objectives, transformation feedback and feedforward, and control.

Systems control

Systems are able to maintain their identity and some form of stability, in changing circumstances, by exhibiting some form of control. Communication of information between components is vital to this control process (see below). Any system which is stabilized by its control mechanisms, and which possesses an identity, can be further understood through its emergent properties. These are properties relating to the whole system which are not necessarily present in any or all of its parts. The patterns of particular systems interactions can determine its emergent properties and in a manner which is not always predictable. For example, safety management only really emerges within an organization through the following properties: visible compliance with operating standards, positive attitudes and shared values, etc. A safety culture may also emerge from such a set of interactions (Turner et al., 1989).

All systems involve control processes that hold them together, otherwise they would degenerate. There are essentially three different levels of control (Carter et al., 1984).

First, the most basic level of control is the self-maintaining causal network. Carter et al. (1984) refer to the natural ecosystems of a tropical rainforest by way of an example. Such a forest can be stable for very long periods of time with 'no sense of purpose, no special controller, no free choice, and no grand design'. Such systems will remain in the same general state indefinitely unless they are radically destabilized by changes too drastic for the network to absorb. Second, there is purposive control. This is more sophisticated in that there are specialized control sub-systems directed towards achieving particular goals. However, these goals are preset – built in – and the level of control reflects 'purpose without choice'. Third, there is purposeful control which reflects 'purpose with choice' and involves deliberate action and free choice – decision making – and a scheme or plan. The 'causal network' model emphasizes systemic factors without allowing for choice and decision making. The 'purposive model' emphasizes the process of pursuing a target and achieving objectives, while the 'purposeful' model emphasizes choice and decision making.

Both purposive and purposeful systems involve special control functions, which often take one of two forms: (a) adaptive control or (b) non-adaptive control.

Adaptive control is sometimes called feedback or closed loop control, because information about the results of the control action is fed back to the control function. The 'loop' is closed and, if necessary, further control actions can be taken in pursuance of the target or objective. Adaptive control is about the active management of systems. Non-adaptive control is control by design. It occurs when effort is put into setting the system up correctly and reliably so that subsequent monitoring and further control actions are not necessary until something goes substantially or dramatically wrong. Non-adaptive control is about the design of systems. To be effective, both adaptive and non-adaptive control need two things: knowledge of the system – a model that lets the control function judge the likely effects of control actions – and awareness of the targets or objectives that the control function is trying to achieve.

In organizations there is usually a hierarchy of sub-systems and an associated hierarchy of control functions. At the lower levels, these control functions are more likely to be purposive and, at the higher levels, purposeful.

Control and purpose: objectives, goals and plans

The preceding discussion of control referred to the pursuit of targets and the achievement of objectives. In organizations, short term objectives, and the tactics required to achieve them, are determined by the practical needs of the local situation. However, local situations often arise out of broader and longer term goals and the strategic plan needed to achieve them. The move from objectives to goals, from the more local to the more global, is often also a move from the more objectively defined to the more subjective. Goals and strategies are often more subjective than objectives and tactics. Goals, in turn, may arise out of even higher level policies (or principles) and associated strategies and these, in their turn, are likely to be even more subjectively defined and begin to reflect a blend of cultural influence and personal attitude. In the organizational literature (see later; also Chapters 5 and 9), terms such as mission, strategy and vision are currently popular. Mission is often used to describe an organization's goals (as opposed to its objectives) while vision is an attempt to capture the ideal outcome of its policies and strategies – where it wants to be.

In summary, so far, control is an important aspect of a system's function, and an expression of its structure. It is intimately related to the achievement of the system's objectives, and involves both control through design and control through management. Understanding failure of control is essential to the analysis of safety in organizations.

Control and safety

The control of an organization is, in itself, an important aspect of that system's structure and behaviour. A breakdown in systems control will inevitably impair the effectiveness of that system and challenge its very integrity and survival.

A breakdown of systems management and adaptive control may threaten the safety of those working within the system. However, such a breakdown is not the only source of threat to the safety of workers. Problems with the design of a system can also generate threats to safety (see, for example, Reason, 1990). Poor design is effectively a failure of non-adaptive control. However, such designs may not even represent failure in the sense that the designers should have known better or done better given current knowledge. What they may reflect is the complexity of the system such that some aspects of systems behaviour or events are unpredictable at the design stage. Unforeseen interactions between system components and activities may 'cause' events which threaten the safety of those working within the system (Groeneweg, 1994). These interactions are what Reason (1990) refers to as 'latent' accidents – disasters waiting to happen.

There are, therefore, at least three ways in which threats to safety can arise within organizations through:

1 a breakdown in systems management and adaptive control;
2 unpredictable events arising from the complexity of the systems design; and
3 failures of systems design and a breakdown in non-adaptive control.

In addition to these aspects of systems design and management, threats to safety may also occur because of the unsafe or 'risky' behaviour of individuals working at all levels within the system. Such behaviour involves both 'errors' and 'violations'. It may be secondary to a failure of systems design or to a breakdown in its management or it may be a reflection of the organizational culture or the interface between the system and its wider environment. The question of organizational culture is discussed later in Chapter 5.

Unsafe or 'risky' behaviours may, for example, be initiated as 'tactics' for achieving valued objectives (personal or organizational) within a poorly designed system. Such violations may not be discouraged and may even be rewarded, deliberately or at least inadvertently, where systems management is poor (Reason *et al.*, 1994). They may be permitted where there is an organizational culture which does not attend to safety, or which does not value actions to protect or develop the workforce. Dissatisfaction and disaffection, which may partly drive violations and also errors, may themselves result from a breakdown or deterioration in the quality of systems control. Finally, unsafe or risky behaviour may be influenced by events in the wider environment– social, economic or political – and to this extent reflect the interface between the system and that wider environment.

Interestingly, these various threats to safety may also represent threats to systems effectiveness and integrity, and organizational survival. It is this possible interaction between the safety (and health) of individual workers and the health of their organizations that has concerned T Cox and his colleagues in the Centre for Organizational Health and Development at the University of Nottingham for some years. The concept of organizational health, as developed at Nottingham, refers to workers' shared perceptions and descriptions of their organization and the way it behaves. It is described in some detail in the next chapter (Chapter 5). From studies on both public and private sector organizations in the UK and elsewhere, a model of what is essentially the subjective organization has emerged. This model posits a number of sub-systems which together form the subjective organization.

In summary, systems theory offers a way of thinking about organizations and the management of safety. It provides the platform and language for exploring the interplay between the individual, the job (tasks) and the organization.

Recent developments in 'systems' thinking

Systems theory is continually evolving. Two recent developments in systems thinking are considered here: first, complexity theory and, second, the design of work organizations through the use of metaphors and the application of the theory of isomorphism. Both are explored and their application to the management of safety briefly discussed.

Complexity theory

Complexity theory is a development of systems theory largely concerned, according to Howarth (1995), with two fundamental questions: first, 'how can complex systems maintain their stability?', and second, 'how do such systems evolve?'. These questions presuppose that systems complexity can be defined, and distinctions can be made between more and less complex systems. Waldrop (1992) argues that a system is complex in the sense that it is comprised of 'a great many independent agents interacting with each other and in a great many ways'. His notion of the complex system applies just as well to the thousands of interdependent components, aspects of hardware and software, and people that together make up an organization, as it does to the quadrillions of chemically reacting proteins, lipids and nucleic acids that make up the living cells of any one of these people. The interplay between the technical and social aspects of organizations ensures their complexity. Complex systems are dynamic and adaptive, but are not necessarily unpredictable (Waldrop, 1992). They are self-organizing and have the ability to bring order and chaos into a special kind of balance at 'the edge of chaos'. Systems at the edge of chaos have enough stability to sustain themselves and enough creativity to evolve in the face of a continual battle between stagnation and anarchy.

The issue of complexity is now at the heart of contemporary systems thinking, and much of the thinking about complexity has been shaped by the Santa Fe Institute founded in the mid-1980s by George Cowan (see Waldrop, 1992). Complexity is now an important issue for a wide range of disciplines and professions that use systems thinking from computer science through physics to biology. It has obvious relevance to the way we think about and manage organizations and the health of those organizations (Howarth, 1995).

Stability of complex systems

Chaos can be created out of even simple deterministic systems but is potentially more common in complex systems. Chaos implies instability. Stability depends on the system having an appropriate structure, and all stable complex systems appear to be aggregations of simpler and more stable components. Stable systems appear to be hierarchies of semi-autonomous components (or modules). Stability appears to be achieved through modularity.

Separate modules are more effective if their activities are coordinated and mutually supportive, and such coordination can only occur at a higher level. One level of coordination creates the simplest form of hierarchical structure. If there are too many modules to be coordinated at this level, chaos may start to develop, and this will lead to modularization at that higher level and the emergence of another even higher level of coordination.

When complex systems exist in complex environments, they will face a variety of challenges and will need to demonstrate different types of integration and coordination between modules. This flexibility of co-ordination requires the development of a hierarchy which can be adapted to pursue a variety of goals adjusting its control structures appropriately. Many organizations achieve this and survive through the development of informal structures which operate alongside the formal and show the necessary adaptability in problem solving to meet the challenges of the external environment.

Schumacher, in his book *Small is Beautiful* (1989), pointed out how much easier it is to work in small groups where everybody knows what everybody else is doing. Contrary to common belief, however, he did not argue that it is impossible to organize things on a large scale; indeed, he developed a theory of large scale organization. This attempted to minimize 'the diseconomies of scale' and was based essentially on the notion of modularity as described here. He also argued that the principle of subsidiarity was important in minimizing diseconomy, delegating as much as possible to the lower levels of the organization. This principle ensures that the hierarchy is of semi-autonomous modules.

The stability and survival of complex systems is possible when they are structured as a hierarchy of relatively simple semi-autonomous modules with a coordination and support function which is adaptable to its needs.

The evolution of complex systems

If simple systems are inherently more stable than complex ones, why is the world dominated by complexity – biological, social and organizational? The answer appears to lie in their evolution and, in turn, in the way systems components (or modules) interact.

Some interactions lead to stable states and some to unstable and chaotic states (as observed above). As already argued, there are, between stability and chaos, intermediate states which have been described as 'the edge of chaos'. It is at 'the edge of chaos' that complexity develops, and complex systems adapt. Such evolution occurs naturally but only when change is possible but not excessive. Optimal adaptation works incrementally and in small steps. Excessive change destroys evolution. The right mixture of stability and change, order and disorder, is necessary. Kauffman (1992) has argued that the evolution of complex systems – their tendency to organize themselves into even more complex systems – naturally occurs at 'the edge of chaos' and is driven by processes of mutation and selection. It occurs in the face of the incessant forces of dissolution described by the second law of thermodynamics. Complexity necessarily leads to a proliferation of semi-autonomous units or modules.

Satisficing

Simon (1981) observed that in any complex system, each constituent module will effectively have its own objectives. He showed that the survival and success of the system depended on it seeking 'good enough' solutions to each of its important objectives. He termed this 'satisficing'. He proved that it was impossible, in economic systems at least, to 'optimize' – to get the best possible solution – for more than one objective, and that concentrating in this way on more than one objective, to the exclusion of all others, can have disastrous results. No successful economic enterprise, according to Simon (1981), pursues profit to the exclusion of all other objectives. There are important messages here for the relationship between profit, quality and safety objectives in any organization. No one objective is more important than any other; no other objective is more important than safety.

Systems design – a metaphorical approach

Complexity places much demand on our ability to conceptualize systems, and various strategies and tools have been developed to aid this process, including the use of metaphors.

A metaphor is defined as the 'application of a name or descriptive term or phrase to an object or action to which it is not literally applicable' (for example, *a glaring error, food for thought, leave no stone unturned*). The use of this idea of 'likeness' as it is employed through metaphor has been successfully applied to several areas of organizational activity (Morgan, 1986; Flood and Jackson, 1991). Flood and Jackson (1991), for example, have described the application of a total systems approach to creative organizational problem solving through the application of systemic metaphors (see Table 4.1). They have identified the following metaphors as filters for looking at problem situations within organizations:

1 the machine metaphor;
2 the organic metaphor;
3 the viable systems metaphor;
4 the cultural metaphor; and finally
5 the political metaphor.

Each of these metaphors has both strengths and weaknesses for the design of effective safety systems. These are further elaborated in Table 4.1.

The machine metaphor, for example, operates within the framework of scientific management (Taylor, 1947) and has most application in the context of the routine and repetitive production of a single item. Within this metaphor, the 'human element' is fitted into the designed structure and

Table 4.1 Application of systemic metaphors to organizational problem solving (adapted from S Cox, 1994)

STRENGTHS	WEAKNESSES
(1) Machine metaphor – typified by theories of scientific management	
• for simple tasks • repetitive production of single product • human element in 'person–machine' fit • stable environment	• reduces adaptability to change • dehumanizes tasks • leads to potential conflict between person/machine
(2) Organic metaphor – typified by humanist views of open system	
• promotes idea of survival and adaptability • focuses on survival needs • stresses feedback for homeostasis • can promote responsiveness to change • change externally penetrated?	• fails to recognize organizations as socially constructed phenomena (what about views from within?) • emphasizes harmony between system elements (what about conflict and coercion?)
(3) Viable systems metaphor – emphasis on active learning and control (Brain Analogy)	
• promotes self-enquiry • encourages creativity • excellent metaphor when there is a high degree of uncertainty • recognition of social constructs	• tends to overlook problem that purpose of 'parts' not consistent with view of 'whole' • usually requires significant change
(4) Cultural metaphor – understood as socially constructed reality (of values and beliefs) that deems certain social practices to be normal, acceptable and desirable	
• shows 'rational' aspects to be culturally dependent • highlights that existing cohesion can both encourage and inhibit organizational development • offers new perspective on organizational change, emphasis on changing perceptions and values of employees	• may lead to explicit ideological control that generates feelings of mistrust • takes time to evolve • can be threatened by political in-fighting • does not give information on how one structures complex in large organizations
(5) Political metaphor – considers relationships between individuals and groups as competitive and involving pursuit of power	
• focuses on key role of power • questions rationality 'whose rationality is being pursued?' • proposes disintegrating strains and tensions (in opposition to 'open-system')	• encourages recognition of individual organizational 'actors' • possibility of over-emphasis on political issues • possibility of generating further mistrust and jeopardising empowerment

employees are expected to passively follow machine-like commands. In practice, this has led to the dehumanization of work and the treatment of the 'person' as a passive respondent in terms of safety outcomes. This has the disadvantage (or weakness) of reducing the individual's potential for active intervention. The machine metaphor precedes 'open systems' thinking in the evolution of management practice.

The organic metaphor involves the concepts of adaptation, change and homeostasis. The organic view is useful when there is effectively an open system operating within a changing environment, and when the readiness to adapt with change is a vital component of survival. This view is said (see Flood and Jackson, 1991) to break down in practice if there is conflict between the various systems components and where change is seen as being externally generated. By contrast, the neurocybernetic or 'viable system' view focuses on active learning and control rather than passive adaptation and dovetails with the theories of isomorphism (see later). This view is one which promotes self-enquiry and encourages creativity. It is most useful when there is a high degree of uncertainty. This approach demands a fully empowered workforce. However, implementation of recommendations based on this metaphor may require considerable and possibly unacceptable levels of change. The cultural and political metaphors are both used in predictable ways; the former may be understood in terms of shared characteristics, whilst the latter looks at the relationships between individuals and groups as essentially competitive and puts a heavy emphasis on the pursuit of power.

Table 4.2 Principles of total systems intervention (TSI) (adapted from Flood and Jackson, 1991)

1	Organizations are inherently complex and cannot be rationalized through one 'management' model
2	Organizational strategies/problems should be addressed against a range of systemic 'metaphors'
3	Systems metaphors may be further linked to established systems methodology to guide intervention
4	Different systems metaphors and methodologies may be used to address separate aspects of organizational activity
5	'SWOT' (Strengths/Weaknesses/Opportunities/Threats) analyses may be facilitated through systems metaphors
6	TSI is an interactive process
7	TSI engages all interested parties during creativity, choice and implementation and review

Flood and Jackson (1991) argue that these five metaphors capture, at a general level, the insights of almost all management and organizational theory (see next chapter). They may be used to analyse and describe problems in combination with a plethora of systems techniques and methodologies (for example, viable systems diagnosis (Beer, 1985), systems analysis (Churchman, 1981), operational research (Ackoff, 1981), sociotechnical systems thinking (S Cox and Tait, 1991), systems engineering and soft systems methodology (Checkland, 1981), and interactive planning (Barstow, 1990), etc.). Such an approach, based on the use of metaphors, has been termed Total Systems Intervention (TSI) and is based on seven principles (Flood and Jackson, 1991). These principles are set out in Table 4.2. TSI is comprised of three phases: creativity, choice and implementation. In the creativity phase, the systems metaphors are used to help 'problem solvers' to think creatively about the organization. The choice phase is about choosing appropriate systems methodologies to suit the particular characteristics of the organization revealed in the creativity phase. The choice of an appropriate methodology has been discussed in a number of management texts (see, for example, Jackson, 1987; Jackson and Keys, 1984). Implementation and review is about translating the preferred 'vision' of the organization, its structure and its general orientation into the recommendations for change.

Organizational isomorphism

Toft and Reynolds (1994) have applied systems concepts to major disasters in organizational systems and have stressed the importance of isomorphism (von Bertalanffy, 1950). 'Organizational isomorphism' postulates that many 'apparently' different systems (in this case, organizational systems) have underlying similarities, possess common properties and, thus, may be liable to common modes of failure. In relation to safety Toft and Reynolds (1994) illustrate the principle of isomorphism by reference to a number of major disasters. For example, they cite the similarity between two fires:

1 the Iroquois Theatre fire, Chicago in December 1903; and
2 the Coconut Grove nightclub fire, Boston in November 1942.

In both cases the decorative fabrics of the interiors were highly flammable and exits were either locked or had not been provided. Both venues were overcrowded, and in neither establishment had the staff been trained to deal with emergencies such as fire (this pattern of circumstances is not unfamiliar in places of entertainment, as subsequent fires have shown). The argument, posited through organizational isomorphism, is that active 'seeking' of systemic failure modes should be extended to failures in other organizations.

Organizational isomorphism thus has particular relevance in proactive systems management (see Chapter 10) where the process of organizational learning extends to both internal or external accidents or incidents.

Systems theory, organizational problems and development

It is argued here that systems theory offers a way of conceptualizing and then solving problems, and has been used in a number of situations and a variety of organizations. It provides a framework for the analysis of problems, and through this analysis, an aid to problem solving. There are several distinct stages in a systems theory approach to problem solving which can be applied to the management of health and safety. These are:

1 the determination of the system's overall function;
2 the identification of the system's constituent parts of functions, and their interrelationships, and its inputs and outputs;
3 the creation of a system's diagram or model or metaphor; and
4 the evaluation and exploration of the system using data from already solved problems, incidents or existing operations.

This systems theory approach is consistent with the systematic problem solving strategy espoused in applied psychology and inherent in current UK and EU health and safety legislation. What the above scheme describes is one of the early stages of problem solving, that of problem analysis (S Cox and Tait, 1991).

Case studies

The development of a model of the subjective organization is one example of the application of systems analysis (see Chapter 9); two other examples are discussed below.

Case study 1: Modelling health and safety management systems
In practice the systems approach to safety management is founded on the need to provide safe and healthy working practices, services and products. Such a goal may be strategically addressed through the design of a fully integrated health and safety management systems model. The first author and her colleagues have worked with a number of organizations (S Cox, 1994; S Cox and Fuller, 1995) to develop safety systems. Figure 4.5 illustrates the model which was developed in 'Organization X' (an organization comprising ten separate units) following an extensive safety management systems audit. The model provides a focus for 'Organization X' and supports key aspects of systems development.

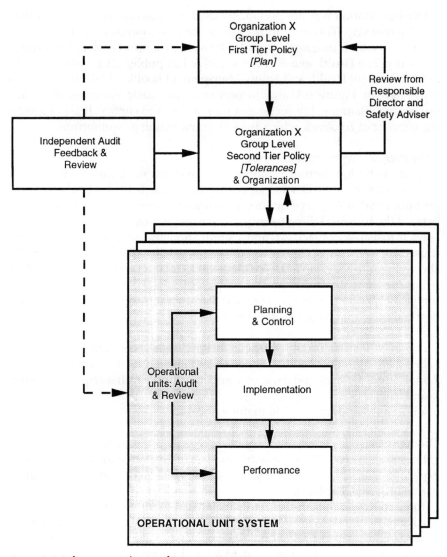

Figure 4.5 Safety systems diagram for organization 'X'

For example, safety management policies and procedures were designed and developed by a Central Policy Unit and were communicated to each of the separate operating units via a policy manual. Feedback on policy implementation was ineffective and occurred on an ad hoc basis. The 'developed' system incorporates feedback on policy implementation as part of adaptive control. Second, since each unit operated semi-autonomously, safety information was not previously used to focus organizational learning (see Chapter 10). This was addressed in the new system's model.

Simple systems diagrams may also be used in support of decision making for safety. The Health and Safety Executive has published a systems model in its Successful Health and Safety Management booklet (HSE, 1991). This is illustrated in Figure 9.1 and incorporates the basic elements of policy, organizing, planning, implementing monitoring and control. It also includes the concept of feedback which is vital in maintaining equilibrium.

Case study 2: Organizational error

Reason (1995) has further elaborated his systems model of organizational error in relation to the Dryden air crash. On 10 March 1989, a Fokker-28 took off from Dryden Airport, Ontario, in heavy snow and crashed one kilometre beyond the runway, killing twenty-one passengers and three members of the crew (Moshansky, 1992). At first examination the accident seemed to be caused by 'pilot error'. However, the inquiry, which lasted twenty months, showed that the captain's flawed decision to take off without the aeroplane being deiced was merely the last chapter in a long history of failures that implicated the greater part of the Canadian air-transport system. These were detailed in a 1600 page inquiry report and included:

1 organizational failings including lack of safety commitment, conflicts between commercial and safety goals and blurred safety responsibilities;
2 individual failings and lack of competence due to inadequate training; and
3 job failings and inadequate maintenance.

The report also identified failures in the regulatory system.

Reason's (1990, 1991) systems model of the aetiology of accidents explicates the Dryden accident in terms of 'active' and 'latent' failure pathways (see Chapter 3).

Toft and Reynolds (1994) have also related systems concepts to safety with particular reference to the concept of feedback with respect to organizational learning and disaster prevention. The principle of organizational isomorphism (see earlier) has been applied to a large number of different disasters including the explosion at Dudgeons Wharf (1969), the Summerland Leisure Centre fire (1973) and the Fairland Home fire (1974).

Summary

This chapter has offered an introduction to systems theory, and some insights into recent developments in systems thinking. It has also attempted to show how such thinking can inform our approach to the management of safety.

Several points are worth reiterating. First, there is an important interplay in systems between structure and function, between organization and behaviour. Second, systems can be very complex, and their complexity can be supported, and the system's integrity and survival sustained, through its structure and control function. Third, control is an important aspect of a systems function, and failure of control can challenge safety in a number of different ways. Fourth, systems can be understood and described through the application of some form of system's analysis. Where the sheer complexity of the system challenges conceptualization, tools such as the use of metaphors are of value.

Overall, systems theory offers an adequate platform on which the following chapters can explore the nature of the organization, its jobs (tasks) and the individual.

References

Ackoff, R.L. (1981). *Creating The Corporate Future*. New York: Wiley.

Barstow, A. (1990). 'On creating opportunity out of conflict: two case studies'. *Systems Practice*, **3**, 339–355.

Beer, S. (1985). *Diagnosing The System For Organisations*. Chichester: John Wiley and Sons.

Carter, R., Martin, J., Mayblin, B. and Munday, M. (1984). *Systems, Management and Change*. London: Harper and Row Ltd.

Checkland, P.B. (1981). *Systems-Thinking, Systems Practice*. Chichester: John Wiley and Sons, Chichester.

Churchman, C.W. (1981). *Thought and Wisdom*. Seaside, CA.: Intersystems Publications.

Cox, S. (1990). *Understanding Human Factors. Course Notes for University of Loughborough Short Course*. October, Loughborough.

Cox, S. (1994). 'The promotion of safe working practices: a systems approach'. In *Proceedings of the 4th Conference on Safety and Well-Being at Work* (A. Cheyne, S. Cox and K. Irving, eds) pp. 40–46, Loughborough: Loughborough University.

Cox, S. and Fuller, C. (1995). *Safety Systems Management Audit Protocol*. Loughborough: Centre for Hazard and Risk Management, Loughborough University of Technology.

Cox, S. and Tait, N.R.S. (1991). *Reliability, Safety and Risk Management: An Integrated Approach*. London: Butterworth-Heinemann.

Cox, S., Janes, W., Walker, D. and Wenham, D. (1995). *Office Health and Safety Handbook*. London: Tolley Publishing Company.

Cox, T., and Griffiths, A.J. (1995). 'The nature and measurement of work stress: theory and practice'. In *The Evaluation of Human Work: A Practical Ergonomics Methodology* (J. Wilson and N. Corlett, eds), London: Taylor and Francis.

Emery, F.E. and Trist, E.L. (1981). 'Sociotechnical systems'. In *Systems Behaviour* (Open Systems Group, eds), London: Harper and Row.

Flood, R.L and Jackson, M.C. (1991). *Creative Problem Solving: Total Systems Intervention*. Chichester: John Wiley and Sons Ltd.

Groeneweg, J. (1994). *Controlling the Controllable*. Leiden University, The Netherlands: DSWO Press.

Health and Safety Executive (1985). *Deadly Maintenance: A Study Of Fatal Accidents At Work*. London: HMSO.

Health and Safety Executive (1987). *Dangerous Maintenance: A Study Of Maintenance Accidents In The Chemical Industry And How To Prevent Them*. London: HMSO.

Health and Safety Executive (1989). *Human Factors In Industrial Safety*. HS(G)48. London: HMSO.

Health and Safety Executive (1991). *Successful Health and Safety Management*. HS(G)65. London: HMSO.

Herbst, P.G. (1954). 'The analysis of social flow systems'. *Human Relations*, **7**, 327–336.

Howarth, C.I. (1995). 'Complexity theory and organisational health'. *International Forum for Organisational Health Newsletter*, August, 9–14.

Jackson, M.C. (1987). 'New directions in management science'. In *New Directions in Management Science* (M.C. Jackson and P. Keys, eds) pp.133–164, Aldershot: Gower.

Jackson, M.C. and Keys, P. (1984). 'Towards a system of system methodologies'. *Journal of the Operational Research Society*, **35**, 473–486.

Kauffman, S.A. (1992). *Origins of Order: Self-Organization and Selection in Evolution*. Oxford: Oxford University Press.

Leavitt, H.J. (1965). 'Applied organizational change in industry: structural, technological and humanistic approaches'. In *Handbook of Organizations* (J.G. March, ed) pp. 1144–1170, Chicago: Rand McNally.

Little, W., Fowler, H.W. and Coulson, J. (1992). *The Shorter Oxford English Dictionary on Historical Principles*, 3rd edn (revised and edited by C.T. Onions). Oxford: Oxford University Press.

Morgan, G. (1986). *Images of organisations*. Beverley Hills: Sage.

Moshansky, V.P. (1992). *Commission of Inquiry into the Air Ontario Crash at Dryden, Ontario*. Ottowa: Ministry of Supply and Services.

Reason, J. (1990). *Human Error*. Cambridge: Cambridge University Press.

Reason, J. (1991). 'Identifying the latent causes of aircraft accidents before and after the event'. In *Proceedings of the 22nd Seminar of the International Society of Air Accident Investigators* pp. 39–46, Stirling: International Society of Air Safety Investigators.

Reason, J. (1995). 'A systems approach to organizational error'. *Ergonomics*, **38**(8), 1708–1721.

Reason, J., Porter, D. and Free, R. (1994). *Bending the Rules: The Varieties, Origins and Management of Safety Violations*. Leiden: Rijks Universiteit.

Schumacher, E.F. (1989). *Small is Beautiful: Economics as if People Mattered*. New York: Harper and Row.

Simon, H. (1981). *Sciences of the Artificial*. Cambridge, MA: MIT Press.

Taylor, F.W. (1947). *Scientific Management*. London: Harper and Row.

Toft, B. and Reynolds, S. (1994). *Learning From Disasters: A Management Approach*. Oxford: Butterworth-Heinemann Ltd.

Turner, B.A., Pidgeon, N., Blockley, D. and Toft, B. (1989). *Safety Culture: Its Importance in Future Risk Management*. Position paper for the Second World Bank Workshop on Safety Control and Risk Management, 6–9 November, Karlstad, Sweden.

von Bertalanffy, L. (1950). 'The theory of open systems in physics and biology'. *Science*, **3**, 23–29.

Waldrop, M.M. (1992). *Complexity*. New York: Simon and Schuster.

Waring, A.E. (1989). *Systems Methods for Managers: A Practical Guide*. Oxford: Blackwell.

5
The organization

Introduction

Most of the employed people in the United Kingdom work within organizations and *think* of themselves as working for organizations. For them, their employer is an organization. At the same time, many of the problems that they report, and which are known to effect their health, satisfaction and behaviour at work, appear to originate in organizational and related management issues (T Cox *et al.*, 1990). An argument has been made out that this might also be true, to some degree, in relation to their safety (Reason, 1995; HSE, 1991). The safety, health and behaviour of employees may depend on the organizational context. It is, therefore, important to examine the nature of organizations, and the relationship between the characteristics of those organizations and the behaviour and safety of their employees. Important mediating factors may be the perceived quality of the organization, its health and also its culture. This chapter explores the organizational context to safety. It uses the ideas introduced in Chapter 4 and provides the essential background to Chapters 9 and 10 which explore the application of organizational thinking to the management of safety. This chapter provides the theory, while Chapters 9 and 10 explore the associated practice. Not surprisingly, it begins by considering the nature of organizations and some of the theories that have been used to understand them.

Organizational theories

Perhaps the first question to answer is 'what is an organization?' and more particularly 'what is a *work* organization?'. Contrasting these two questions immediately implies the important point that not all organizations are work organizations, and that whichever definition of an organization is adopted, it must be capable of modification to clearly distinguish between those organizations which are work organizations and those which are not, such as 'social clubs' or 'voluntary organizations'. In answering these questions, researchers have used a number of levels of analysis (see Blau, 1957). Three are clearly identifiable in the relevant literature; they focus on either: (1) individual behaviour within organizations; (2) the characteristics and function of particular aspects of organizational structure; or (3) the characteristics and behaviour of the organization itself as a collective entity (Scott, 1981). Here the authors are primarily concerned to explore individual behaviour within the context of the organization. The characteristics and behaviour of the organization itself provide the *context* or 'environment' within which, individual behaviour is enacted. What the authors are attempting is an explanation of how such a context determines and, at the same time, reflects the attitudes, behaviour and safety of employees. There are several different types of explanation available. These have been classified in terms of their historical development and the types of theory that they reflect: first classical organization theory, then the human relations school, and, most recently, systems theory (see Bryans and Cronin, 1983). Arguably, the latter theories – the human relations school and systems theory – have proved the most influential in the present context. Theories within the human relations school have often been labelled *social psychological* and are exemplified by the writings of March and Simon (1958), Porter *et al.* (1975), and Duncan (1981). This chapter draws on both the social psychological and systems theory perspectives.

Organizations

In their book, *Organizational Behaviour*, Buchanan and Huczynski (1985) suggest that 'organizations are social arrangements for the controlled performance of collective goals'. There are three important elements to such a definition. The first is that organizations are social arrangements, and are thus represented in the existence and behaviour of a group, or groups, of people. Organizations are about the interactions between individuals in those groups and between such groups. People and groups are key components in the organization as a system. The second important element is that organizations are concerned with achieving collective goals. This implies that organizations, like all systems, exist for a purpose and that purpose is represented in those collective goals. It is an interesting question whether or not most, if not all, members of organizations know and understand, if not share, those collective goals. This question might provide one dimension whereby the health of the organization could be assessed. The third important element is that organizations function through controlled performance in pursuit of those collective goals. This introduces the notion of management and with it that of some division of labour. The control exercised by management is usually both adaptive and purposeful (see Chapter 4).

A somewhat similar social psychological definition has been offered by Duncan (1981): organizations are 'a collection of interacting and inter-dependent individuals who work towards common goals and whose relationships are determined according to a certain structure'. Duncan's (1981) definition emphasizes the interaction and interdependence of the individuals who make up the social groups, which in turn form the organization, and introduces the idea that organization structure, as well as its management function, plays a part in controlling individual behaviour. This point is explored in more detail in Chapter 10.

Buchanan and Huczynski's (1985) and Duncan's (1981) approaches to defining organizations draw attention to their reality as *social constructions*, and this may be contrasted with a more physical view of organizations as embodied in their site, plant, machinery and specified work systems and procedures and hierarchies. Thus, we might be able to contrast the social organization – the organization as people – with the technical organization – the organization as hardware and software – and ask to what extent these different aspects of the organization are consistent. Does the technical organization support and facilitate the social organization and vice versa? As before, questions such as this may offer dimensions whereby the health of organizations might be assessed. With specific regard to health and safety, the importance of the social organization – people – and the need to manage and control it properly have been clearly recognized by the UK Health and Safety Executive (HSE) (1981):

> To prevent accidents to people and damage to plants and the environment, one needs to ask how management should be involved. Management's responsibility is to *control* work – both its human and its physical elements, and accidents are caused by failures of control. They are not, as is so often believed, the result of straightforward failures of technology: social, organizational and technical problems interact to produce them (HSE, 1981, p.6).

In summary, social psychological definitions of organizations treat them as social constructions, systems in which the main components are people and groups of people. Such organizations exist for the controlled achievement of shared goals. The control exercised within these organizations is partly through structure, and partly through management and management systems. The latter has been said to be both adaptive and purposeful; the former is control by design. It is recognized here that the social organization has to be considered in relation to the technical organization; these are complementary and interdependent aspects of the total organizational system. Any adequate theory of the organization as a total system has to embrace both aspects; one attempt to do so, sociotechnical theory, is described below. The need to work with such theories is made obvious by reference to reports on recent safety disasters which consistently point up the interplay of technical, social and organizational factors in their aetiology (see HSE, 1981, above).

Organizations as sociotechnical systems

In Chapter 4, open and closed systems were described and contrasted. Many early theorists tended to view the organization as a closed system (see Emery and Trist, 1981) despite the implications of such an approach. For example, the inherent tendency of closed systems is to grow towards maximum homogeneity of their parts with steady states only achieved by the cessation of all activities. Such theorists believed that organizations could be sufficiently independent to allow most of their problems to be analysed with reference to their internal structures and without reference to the external environment in which they exist. The alternative approach is to consider organizations as open systems.

It was argued in Chapter 4 that open systems depend on exchanges with their wider environments to survive; they import things (materials, information, etc.) across their boundary with that environment, transform those imports and then export things back out to that environment. Open systems are self-regulating and adaptable. They grow through a process of internal elaboration (Herbst, 1954) and may spontaneously reorganize towards states of greater complexity and heterogeneity (Waldrop, 1989). They manage to achieve a steady state while working – a dynamic equilibrium – in which the organization as a whole remains constant with a continuous throughput and despite a considerable range of external changes (Lewin, 1951). This approach to organizations would seem more appropriate than the closed systems model, and, indeed, current systems thinking applied to organizations treats them as if they are, to a large extent, open systems (Emery and Trist, 1981; Buchanan and Huczynski, 1985).

The open systems view of organizations has also been termed the 'organic analogy' (see, for example, Rice, 1963; Miller and Rice, 1967; Flood and Jackson, 1991) because it implies that organizations have properties in common with living organisms, including the capacity for growth (and contraction). The concept of organizational health draws directly on this analogy in terms of a model derived from the physiological mechanisms underpinning individual health (T Cox and Howarth, 1990).

The notion of the organization as an open system is appealing because the introduction of systems theory allows the development of a total systems view, but in itself it is not fully compatible with the essentially social psychological definitions of organizations discussed above. Some further reconciliation of these approaches was always required, and the integration of these two sets of ideas produced sociotechnical systems theory (see Figure 5.1). The development of sociotechnical thinking is largely attributed to the work of Emery and Trist and their colleagues (for example, Trist and Bamforth, 1951; Emery and Trist, 1960; Trist *et al.*, 1963; Emery and Trist, 1981).

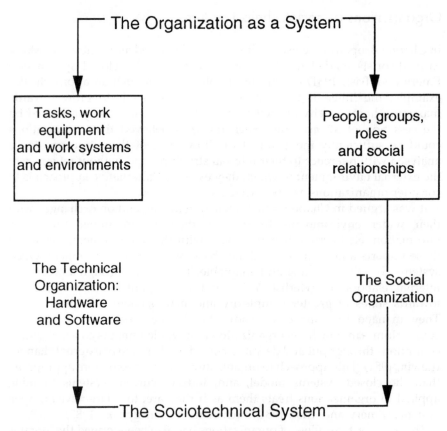

Figure 5.1 Sociotechnical systems theory

The concept of the organization as a sociotechnical system derives from the fact that all work systems require both a technology (hardware and software: tasks, equipment, work systems and tasks, and environments) and a social organization of the people who use that technology. These different components of the organization are, at the same time, independent and interdependent. For example, in their studies in the coal mines around Durham, UK, and in a textile mill in Ahmedaba, North West India, Trist and his colleagues (1951, 1963) showed, at least in these cases, that the social organization has social and psychological properties that are not dependent on the demands of the technology used. Findings like theirs make it clear that sociotechnical systems theory is not simply a technological determinist approach in the mould of those espoused by authors such as Woodward (1965) and Perrow (1967).

Sociotechnical theory seeks to describe the nature of the technical and social components of organizations, analyse their interdependence and design the 'best fit' between them. Theorists argue that organizations must be designed in such a way that the needs of each component are met to some extent and that they are consistent one with the other. As suggested earlier, the extent to which these two aspects of the organization are consistent and mutually supportive can be treated as a measure of the health of that organization. An unhealthy organization is one, for example, where the technical system has been designed without taking into account the needs of the social system: under such circumstances the system as a whole will not work effectively. Interestingly, Trist and his colleagues (1951, 1963) have argued that an effective sociotechnical system design cannot completely satisfy the needs of either the technical or the social systems: this sub-optimization appears to be a necessary feature of sociotechnical design. Such a strategy was referred to as 'satisficing' in Chapter 4.

In their studies in Durham, Trist and his colleagues (Trist and Murray, 1948; Trist and Bamforth, 1951) compared what was then a conventional form of work organization for mining – the conventional longwall system – with a composite system effectively based on group working. The conventional system combined a complex formal structure with simple work roles: miners worked a single part task, had only limited interactions with others in their particular task group, and were sharply divided from those outside that group. The composite system, in contrast, combined a simple formal structure with complex work roles: miners performed multiple tasks with a commitment to the overall group task, had much contact with others in the group and were involved in the self-regulation of the group. The two different systems worked with the same technology and the same coal seam; however, their social systems and the miners' task profiles were very different. The composite group system represented a sociotechnical design which, incidentally, had embraced the principle of 'subsidiarity' – control at the lowest appropriate level – espoused in Chapter 4.

The available data suggested that in terms of safety and health, the composite 'group' system was the superior. The absenteeism data (expressed as a percentage of possible shifts) clearly indicated this superiority with 20 per cent absenteeism recorded for the conventional system and only 8.2 per cent recorded for the composite group system. The figures included those for absenteeism due to accidents which also reflected the superiority of the composite group system over the conventional system:

6.8 per cent for the conventional system versus 3.2 per cent for the composite group system.

The work of Trist and his colleagues (for example, Trist and Bamforth, 1951; Emery and Trist, 1960; Trist *et al.*, 1963), and other researchers based in the Tavistock Institute, London (for example, Herbst, 1962), led to the following conclusions, that:

1 work in groups is more likely to provide meaningful work, develop responsibility and satisfy human needs than work that is allocated to separately supervised individuals; and
2 work can be organized in this way regardless of technology.

However, Emery and Trist (1981) have emphasized that their findings did not suggest that work group autonomy should be maximized in all organizational settings. They argued that there is an optimal level of grouping which needs to be determined in relation to the requirements of the technical system but that the relationship between level of technology and level of grouping is not a simple one. They also suggested that the psychological needs that are met by grouping are *not* employees' needs for friendship while working (as the Human Relations School might suggest):

> Grouping produces its main psychological effects when it leads to a system of work roles such that the workers are primarily related to each other by way of the requirements of task performance and task interdependence. When this task orientation is established the worker should find that he (or she) has an adequate range of mutually supportive roles. As the role system becomes more mature and integrated, it becomes easier for a worker to understand and appreciate his (or her) relation to the group (Emery and Trist, 1981, p. 174).

Despite these caveats, the work of sociotechnical systems theorists has led to the general promotion of autonomous work groups as a form of work organization. The implementation of such groups has been viewed as a form of job enrichment but one applied not to individuals but to work groups. Autonomous group working can be effectively applied to groups of people whose work is related or interdependent. Buchanan (1979) has attempted to set out what such work group organization should involve:

1 job rotation or physical proximity when individual tasks are: interdependent, stressful and lack a perceivable contribution to the end product;
2 grouping of interdependent jobs to give: whole tasks which contribute to the end product, control over work standards and feedback of results, and control over boundary tasks;
3 communication channels; and
4 promotion channels.

Systems theory approaches to organizations have been criticized for treating the organization as if it were an entity separate from the individuals who are part of it, and for ascribing goals and actions to it rather than to those individuals (see Beazley, 1983). Sociotechnical systems theory overcomes this criticism to some extent, making explicit the role and importance of the social component of the organizational system. However, the emphasis is still not strictly on the individual, as the notions of work group and work roles are arguably more important for such theorists than that of the individual employee. Whatever the exact nature of the emphasis in systems theory approaches to organizations, the very nature of the theory lends itself to the analysis of organizational level problems and actions, and can lead to a variety of organizational strategies for change, in general, and problem solutions in particular (see Chapters 9 and 10).

Systems thinking has been variously and successfully applied to a variety of organizational issues during the last thirty years (see, for example, Miller and Rice, 1967; Scott and Mitchell, 1972; Open Systems Group, 1981; Beer, 1985; Waring, 1989). Furthermore, recent developments in systems thinking, such as the interest in complexity theory (see Chapter 4), are now beginning to influence organizational thinking (for example, Howarth, 1995). Among a plethora of relatively new concepts is that of organizational health. This has been offered in several different forms; many of them rather unimaginative and not novel. What is discussed here is the concept as developed in the Centre for Organizational Health and Development, at Nottingham, by the second author and his colleagues.

Organizational health

The concept of organizational health is based on an analogy with individual health, and is a derivation of sociotechnical systems thinking. It is about the nature and viability of organizations as systems, and includes measures of the social organization and its relationships with the technical organization.

T Cox and Howarth (1990) presented a model of the health of an organization built upon an analogy with an individual's physical health. The model places particular emphasis on the sub-systems[1] which together describe the internal functioning of the organization. These sub-systems relate to the total organization in much the same way as physiological systems, such as the cardiovascular or respiratory systems, relate to the whole person.

---

[1] In previous publications, these sub-systems have been referred to as *psychosocial* as a convenient shorthand. They involve psychological, social and software aspects of the organization: people, groups, procedures and systems.

In a study of education provision in Britain, the second author and his colleagues (see T Cox *et al.*, 1992) identified, through teachers' descriptions of their schools, three possible sub-systems: these represented the school in terms of task completion, problem solving and staff development. They found that the quality of the task and problem solving sub-systems could be related to teachers' experience of work-related stress and their report of general symptoms of ill-health, and also to absence behaviour and declared intentions to leave. Unfortunately, in this study no measures of safety performance were available. However, the study did demonstrate that the health of employees and (some aspects) of their work behaviour were related to the health of the organization (school) in which they worked. Such employees, it was generally argued by T Cox and Howarth (1990), function more effectively when (1) they have a coherent and positive perception of the organization, and (2) this perception is largely congruent with organizational reality. The coherence which exists within their perceptions of the organization, between its various sub-systems, was said to reflect at least two factors: the effectiveness of its *communication* systems, and the strength of its *social structure*.

Organizational sub-systems

T Cox and Leiter (1992) have offered a preliminary commentary on the nature of three sub-systems described through employee perceptions.

The absence of support at the level of primary *task completion* for strongly espoused organizational goals and objectives may reflect unresolved conflicts regarding policy throughout the organization. Argyris (1982) proposed that such contradictions of an organization's espoused goals and objectives and its actual practice were signs of fundamental flaws in organizational functioning. The strain of such contradictions may undermine organizational effectiveness by increasing unproductive demands on staff while simultaneously undermining their resource base, thereby increasing their potential for experiencing work-related stress (see T Cox and Griffiths, 1995).

The nature and effectiveness of *problem solving* processes are sensitive indicators of organizational health. Problem solving is the occupational activity which offers the widest scope to the creative competence and professional judgement of employees both as a group and as individuals. Problem solving conducted solely as an individual responsibility or under pressure or hampered by arbitrary constraints denies the value to the organization of staff competence. This loss will be exacerbated when the target of ineffective problem solving is the organization itself. By undermining efforts to improve organizational functioning, poor problem solving diminishes hope for future improvement. It encourages the perception of the organization as stagnant with little potential for growth. The thoroughness of problem solving procedures, formal and informal, and the broadness of staff participation in those processes are defining characteristics of an organization's culture (Janis and Mann, 1977).

Employees often judge an organization's potential for growth through their own aspirations for personal and professional *development*. The thwarting of these aspirations can undermine staff confidence in the organization's future (Hendry, 1991).

Harmonious social relationships may reflect *consensus* regarding major goals and objectives, as well as a *balance* between those goals and objectives and the availability of appropriate resources. An organization characterized by harmonious social relationships may generate less stress for its members. The associated consistency in goals and values may facilitate the setting of priorities, thereby reducing conflict between organizational members regarding the provision of resources (Hackman, 1986). Similarly, consistency may make it easier for individuals to effectively allocate their time and energies. In contrast, organizations lacking such congruence place increasing pressures on their members (Holland, 1985).

Assessing organizational healthiness

The Organizational Health Questionnaire (OHQ) is being developed to help researchers and managers assess the health of organizations. The OHQ is the intellectual property of the Centre for Organizational Health and Development, Department of Psychology, University of Nottingham (T Cox *et al.*, unpublished). In its present form, it consists of three scales measuring:

1 the goodness of the organization as an environment for task completion;
2 the goodness of the organization as a problem solving environment; and
3 the goodness of the organization as a development environment.

The OHQ also asks for information on staff absence, the impact of absence, and staff turnover and intention to leave.

The OHQ captures employees' perceptions and descriptions of the organization. These data are subjective but acceptable given that those employees have 'expert' knowledge of their organization by reason of their experience of working within it and, necessarily, of their reflections on that experience. The concept of the 'expert' or 'reflective' employee is important. Experience gained unthinkingly – without any form of awareness or critical reflection – may not be sufficient for the present purpose. Fortunately, while the level of critical reflection must vary from one type of work, workplace or individual employee to another, it must be rare for an employee never to think about or discuss their work with another, or not to have received some form of instruction or training on how to do it. As a result, most employees' perceptions and descriptions of their work and work organizations as systems have some validity as expert evidence or data. Furthermore, the full extent of the validity of a systems description derived from conversations with workers can be tested empirically.

As already stated, the OHQ was originally developed through a series of studies involving teachers in UK schools (T Cox *et al.*, 1988, 1989; T Cox *et al.*, 1992). These studies have been extended in two ways. First, the instruments utility was tested in Singaporean and Taiwanese contexts. Second, its applicability in the UK was examined in relation to healthcare and manufacturing organizations. In both cases, confirmatory analyses demonstrated the replicability of the original structure.

The health of organizations, it has been suggested earlier, is the reflection of several different levels of organizational analysis. At the highest levels, the health of the organization is a reflection of how well matched the technical and social organizations are and how realistic employees' perceptions of it are. At the lowest level, it is a reflection of the reported goodness of one or all of its sub-systems. At the intermediate level of analysis, health is reflected in the degree of integration of the three sub-systems. This type of thinking about organizational health is based on a 'level of description' model and implies the possible existence of a series of measures varying in 'granularity'.

Applications to safety

Recent research has suggested that measures of health at the finest level of granularity – the goodness of the systems components – are relatively strong predictors of outcome measures such as workers' experience of stress, their self-reported wellbeing and absence behaviour. There is, therefore, evidence that the health of the organization is related to that of its employees. There are no data, yet, which relate organizational health to more direct markers of worker safety or the incidence of accidents. However, there is room for informed speculation.

It has been argued in Chapter 4 that there are at least three ways in which threats to employee safety can arise within systems; through a breakdown in systems management and adaptive control; through unpredictable events arising from the complexity of the system; and through failures of systems design. In addition to these aspects of systems design, behaviour and management, threats to safety may occur because of unsafe or 'risky' behaviour at all levels of employment in the system: errors and violations. Such behaviour may be secondary to a failure of systems design or a

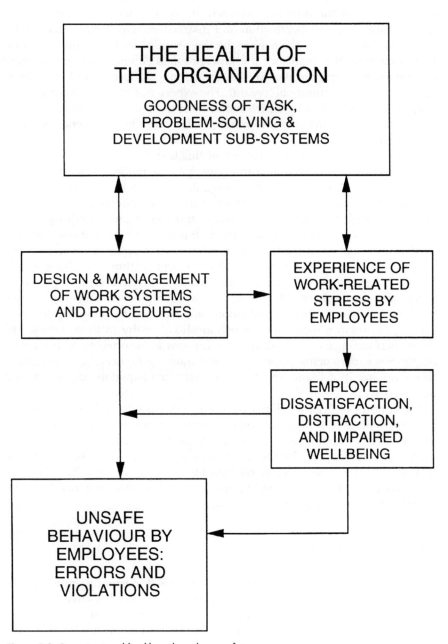

Figure 5.2 Organizational health and employee safety

breakdown in adaptive control, or it may be a reflection of the organizational culture or the interface between the system and its wider environment. Dissatisfaction, disaffection and distraction, which may partly drive unsafe behaviour, may themselves be the result of an unhealthy organization. Therefore, the quality of the task, problem solving and development environments may give rise to dissatisfaction, disaffection and distraction and, in turn, drive unsafe behaviour. The experience of stress by employees may be the mechanism by which their perceptions and cognitions about the organization are translated into unsafe behaviour. The experience of work-related stress has been explored by T Cox (1978, 1993; T Cox and Griffiths, 1995) and is the subject of Chapter 8 of this text.

This argument is represented in Figure 5.2. The model outlined in Figure 5.2 suggests that the health of an organization can affect safety both (1) through the design and management of its work systems and procedures and (2) through the experience of stress and the organization's impact on employee health and behaviour at work. It is argued here that the latter can moderate the effects of the former on safety. The interplay between the two is represented in a true interaction. Healthy organizations are those which, among other things, both design and manage effectively safe systems of work, and also enhance the health of their employees encouraging safe working behaviour. Such organizations are expected to demonstrate better safety performance than those which are less healthy in these terms. The safety performance of unhealthy organizations is expected to be poor with employees experiencing poor health and showing reduced commitment as those organizations attempt to implement inadequately designed and managed work systems.

The concept of organizational health is very close to and, in some ways, overlaps with that of organizational culture. It is possible that they are two different views of the same set of phenomena, the conceptual difference being generated largely by the differences in methodology associated with their study. The issue of their relationship is not resolved. The following sections discuss what is meant by organizational culture, in general, and by safety culture in particular.

Organizational culture

The general concept of organizational culture is currently important in theories of organizations, and the more particular concept of safety culture is equally important for our understanding and management of safety at work. In a statement following the public inquiry into plans to build a new pressurized water nuclear reactor at Sizewell in Essex (UK), a government minister discounted fears that the reactor could be affected by an accident similar to that which occurred at Chernobyl *because* the UK nuclear industry had a 'superior safety culture' (Ministerial Statement, 1987, p. 36). Since the Chernobyl disaster, the development of an 'appropriate' safety culture has been seen within the Western nuclear industry as an important human factor requirement of nuclear operator training (Broadbent, 1989). The concept of a safety culture has also informed other inquiries and analyses of safety failures and related disasters, such as the Clapham Junction rail disaster in London. Here, the public inquiry found a poor safety culture within British Rail to be an important determinant of the accident (Hidden, 1989).

Given the importance of the culture concept to organizational theory and its particular relevance to safety, it is not a trivial question to ask 'what is meant by the term "organizational culture" and how does such a term apply to safety in organizations?'.

The general concept of culture has been widely used throughout social science and a multiplicity of definitions are available. Traditionally, there have been two principal models: that which defines culture in terms of *behaviour* and that which defines it in terms of *meaning*. Rohner (1984) states that:

> there are, for example, those who view culture as being behaviour; the regularity occurring, organized modes of behaviour in technological, economic, religious, political, familial and other institutional domains within a population. In contrast to the various 'behavioural' modes of culture are a group of theorists who hold that culture is a symbol system, an ideational system, a rule system, a cognitive system, or, in short, a system of meanings in the heads of multiple individuals within a population (Rohner, 1984, p.113).

In terms of organizational theory – and also safety science – it is the second type of definition that has held greatest sway. For example, organizational culture has been defined by Schein (1985) in terms of employees shared values and perceptions of the organizations, beliefs about it, and common ways of solving problems within the organization. As such, it must have relative stability and not change on an hourly, daily or even weekly basis. Such a definition is very close to the Nottingham concept of organizational health. In much the same vein, Turner *et al.* (1989) have defined organizational culture in specific relation to safety as:

> the set of beliefs, norms, attitudes, roles, and social and technical practices that are concerned with minimizing the exposure of employees, managers, customers and members of the public to conditions considered dangerous or injurious (Turner *et al.*, 1989).

Within this broad definition, Pidgeon (1991) argues that safety culture can be conceived of as the constructed systems of *meanings* through which a given person or group understand the hazards of their world. Such a constructed meaning system specifies what is important and legitimate to them and explains their relationship to matters of life and death, work and danger.

While there is apparent widespread agreement on these types of definition, they are broadbased and there must be a worry that they will therefore become a catch-all for social psychological and human factor issues. The broadness of these definitions weakens their scientific utility especially as much of the development of such definitions has been theoretical and not empirical. While definitions derived from empirical studies would undoubtedly appear narrower and less 'rich' in meaning, they would have the advantage of being testable, and data might be provided on their reliability, validity and degree of bias.

Schein's (1985) definition was derived from his studies of, and observations and reflections on, organizations. It produces a model of culture which is somewhat similar in structure to the model of organizational health described in this text and derived from empirical studies on schools and healthcare organizations. The concept of organizational health is then closely related to that of organizational culture (as defined by Schein, 1985) and captures some aspects of culture, but probably not all. At the same time, it is an inherently evaluative concept, while that of culture is essentially neutral at that point of definition. However, although the general concept of culture should be neutral at the point of definition, Pidgeon (1991) argues that the use of the term 'safety culture' has an implicit normative element, and thus one might ask questions about the characteristics of 'good' and 'poor' safety cultures. This question is addressed below.

In terms of systems theory, organizational culture can be treated as an emergent property of the organization as a system. That is culture is a property of the whole system, a reflection of the interaction between its individual components and processes. It is a reflection of the state and a function of those individual components and processes, and their interactions, and it influences them, but it is not located in any single or particular component, process or interaction. It is a gestalt: it resides in the sum of its parts and not in any one of them. This observation may offer a further differentiation of the concepts of organizational culture and health. The former is an emergent property of the organization as a system, while the latter is a measure of the quality of that system, or of the compatibility of its components.

Any system can be deconstructed into its component sub-systems and many, if not all of them, might be treated as systems in their own right. Thus each sub-system has the potential for a culture, and just as these systems and sub-systems may be hierarchically arranged and reflect different organizational structures and functions, so might the emergent cultures and sub-cultures. As systems have sub-systems, cultures have sub-cultures. Indeed,

Adams and Ingersoll (1989) have commented that it is often best to discuss organizational culture in terms of constituent sub-cultures. In this vein, it has been variously argued that organizational cultures and sub-cultures are nested (for example, Pidgeon, 1991) and undoubtedly overlapping being mutually influential across and between levels in that nested structure. In theory, one might expect a safety culture to exist as an emergent property of the safety system.

Characteristics of a 'good' safety culture

There is an obvious lack of empirical data to provide a secure answer to the question of 'what are the characteristics of a good safety culture?'. There have been few studies which have been broad enough in their design to capture, in one attempt, all the potential aspects of safety culture implied in its definition. Most studies have focused on one or two of its more fundamental characteristics.

Pidgeon (1991) has implied that a good safety culture has, in essence, three main dimensions: (1) norms and rules for effectively handling hazards; (2) positive attitudes towards safety; and (3) the capacity for reflection on safety practice (reflexivity). One may, of course, question whether, given the breadth of the definition offered by Turner et al. (1989), these three dimensions suffice. However, setting that question aside, it is obvious that most of what is being covered by these three dimensions should be related to social psychological processes and, in particular, social cognition and the social construction of reality. The norms and rules referred to must be resident in the shared perceptions and cognitions of employees, must be a reflection of their beliefs and values, and not simply published organizational guidelines, bylaws and instructions. Equally, the capacity for reflection must again be resident in social processes and cognitions rather than in formal procedures, meetings and workshops. If this point is not taken strongly enough then the utility of safety culture as a 'separate and valid' concept is lost. While this point has to be made in relation to Pidgeon's (1991) first and third dimensions, the second – that of attitudes to safety – is by its very nature immune to this problem. It is perhaps because attitudes to safety sit the most comfortably of all three dimensions in the safety culture model that they are often treated as that model's defining characteristic. Positive attitudes to safety are arguably the most important aspects of a 'good' safety culture. For a discussion of attitudes to safety refer to Chapters 7 and 12.

Although there have been few adequate studies of good safety cultures, in the sense that they have tapped all aspects of its definition (see Pidgeon, 1991), other studies have shed light on this issue. In particular, studies of 'safe' organizations – those with a good safety performance – are illuminating. For example, comparative studies between high and low accident plants have revealed some relevant results (Advisory Committee for Safety in Nuclear Installations (ACSNI), 1993); these are discussed later.

One of the early studies on the correlates of safety performance was carried out in a tractor assembly plant by Keenan *et al*. (1951). They analysed a total of 1941 lost time accidents over a five-year period and identified all the possible factors contributing to accident causation. Correlational analysis across forty-four departments, based on independent scores on each of the previously identified causation factors and accident rates, revealed that a 'clean and comfortable working environment' was the most significant predictor of 'good' safety performance. This may seem an obvious outcome considering the nature of the operation – the processes involved foundry work – and the concomitant hazards. However, this relationship still held when the effects of other variables (for example, operational congestion, constant production pressure, manual effort, etc.) were accounted for using partial correlation techniques. Another, and in some ways more interesting, finding from this early study (Keenan *et al.*, 1951) showed that the greater the 'promotion probability' in a department the lower the accident rate, assuming that comfort of the environment was held constant. Finally, predictably Keenan *et al*. (1951) found a relationship between 'obvious dangers' and high accident rates. However, further analysis of the incidence of these accidents showed them to be somewhat unrelated to the actual source of danger. This finding may reflect some of the cognitive and perceptual issues outlined in earlier sections, where the behaviour of individuals is driven by their heightened awareness of certain hazards and risks. Keenan *et al*. (1951) refer to 'an almost hypnotic' effect in their description of this finding. Keenan *et al.*'s (1951) findings are not inconsistent with the theory of organizational health, and their first two findings (as described here) can be said to be expressions of the importance of the task and development environments. To a certain extent, they might also be a reflection of a good safety culture.

More recent comparative studies have included DeMichiei *et al.*'s (1982) and Peters' (1989) work in the coalmining industry; Cohen *et al.*'s (1975) studies in lumber, metals and machine manufacturing; Smith *et al.*'s (1978) studies across a range of industries and Zohar's (1980) seminal work on safety climates. Lee (1993 and 1995) has summarized the key characteristics of low accident plants based on the evidence of these studies (see Table 5.1). These include having effective communication at all levels; showing evidence of organizational learning; a strong focus on safety and senior management commitment; effective and participative leadership; quality safety training which incorporates skills training; clean and comfortable (relative to the task) work environments; high levels of job satisfaction; and a workforce composition which recruits, rewards and (thus) retains employees who work safely and have lower turnover and absenteeism (as distinct from higher productivity). Many of these characteristics directly

Table 5.1 Characteristics of low accident plants (adapted from Lee, 1995)

- A high level of communication between and within levels of the organization; less formal and more frequent exchanges; safety matters are discussed; managers do more walkabouts

- Good organizational learning, where organisations are tuned to identify necessary changes

- A strong focus on safety by the organization and its members

- A senior management that is committed to safety, giving it high priority, devoting resources to it and actively promoting it

- A management leadership style that is cooperative, participative and humanistic, as distinct from autocratic and adversarial

- High level of quality training, not only specifically on safety, but also with safety aspects emphasized in skills training

- Clean and comfortable (relative to the task) working conditions; good housekeeping

- High job satisfaction, with favourable perceptions of the fairness of promotion, layoff and employee benefits as well as task satisfaction

- A work-force composition that often includes employees who are recruited or retained because they work safely and have lower turnover and absenteeism, as distinct from higher productivity

---

relate to Pidgeon's (1991) three dimensions of safety culture and, arguably, can be said to derive, in part, from the three environments described by the theory of organizational health.

Organizational culture and safety

The question is how and to what extent can the concept of safety culture be used to explain and predict safety in the form of accidents at work? An organization's safety culture must also show a relative stability. Given this stability, how can the concept of safety culture be used to explain and predict accidents at work which are deviations from a stable and safe work system?

Accidents are, in most cases, relatively low frequency events, and, although the product of a number of interacting processes, are discrete events in themselves. How can a relatively stable and chronic feature of organizations, such as its safety culture, be used to explain individual accidents – these low frequency discrete events? The answer is that, in these terms, it cannot. However, this does not mean that the concept of safety culture is irrelevant to safety research: it does mean that the way in which it is invoked has to be carefully thought through.

The solution

In order to show an association between two variables, such as safety culture and accidents, there has to be some variation in both, and, logically, that variation has to have a similar basis in time.

Our measure of safety culture has to show some variance despite its stability. This can be achieved in, at least, three different ways. First, culture as measured can be treated as a *between organization* variable, and the culture of different organizations contrasted and then related to differences in their accident rates. Second, culture as measured can be treated as a *within organization* factor but different parts or divisions within the organization can be contrasted. This second paradigm is the within–between option. The third approach is again to treat culture as a within organization variable but to capitalize on *between employee* differences in perceptions or reporting of safety culture. At this level and in this way, culture as measured can be related to individual measures reflecting accident rates or safety behaviour.

Given that there are these three approaches to the measurement of safety culture, then these will determine how an accident rate has to be conceptualized and measured. First, this means that, above all, that measure must be grounded in a time scale compatible with that for safety culture. We must ask how stable is the measure of culture: over what period of time would it be naturally expected to change? Perhaps it has been estimated that a sensible half-life for safety culture would be six months, then this period would be used as the time window for the measurement of accident rate; the number of accidents occurring in a six-month period would be measured. If very few if any accidents happened in this period, then it will prove difficult to use the accident rate as a variable, and other accident relevant measures – such as near misses – would have to be explored.

Second, this means that the measure of accident rate must be taken at the same level as that of culture. In the first approach described above, the cultures of, say, two organizations would be contrasted. The measure of accident rate which would be related to this contrast, must also be at the between organization level. Thus, in this example, the accident rate for the whole organization would be determined over a six-month period. In the second example, the contrast is within an organization but between its parts. In this example, accident rates would be determined for each of the parts being contrasted. In the final example, the contrast is between different employees' perceptions of organizational culture. Accident rates would then have to be determined for each individual over the six-month period. Of course, these are the base data, and can be aggregated variously to support the 'higher' level analyses referred to above.

What is being attempted here is a scientifically derived answer to the question which relates safety culture and accident occurrence. Other approaches are possible, although not adequate. For example, it would be possible to produce an 'expert' analysis of the single accident which invokes culture as an explanatory variable and which does not draw on any empirical evidence. Such analyses are most usually unsatisfactory, and where tested, are usually invalid and unsupported.

Summary

The nature of work organizations has been explored in this chapter through the combination of social psychological and systems theory approaches. The latter was employed as a vehicle, among other things, for reconciling the notion of the social organization with that of the technical organization. This reconciliation was illustrated by consideration of sociotechnical theory. Two currently important concepts within most organizational theories are those of organizational health and culture. These were discussed and the overlap between them explored. The implications of both for safety and safety research were discussed.

One of the main points which is implicit in the information and arguments offered in this chapter is that knowledge of what organizations are and how they work is essential for the effective management of safety.

References

Adams, G.B. and Ingersoll, V.H. (1989). 'Painting over old works: the culture of organisations in an age of technical rationality'. In *Organisational Symbolism* (B.A. Turner, ed), Berlin: Walther De Gruyter.

Advisory Committee for Safety in Nuclear Installations (ACSNI) (1993). *Organising for Safety – Third Report of the Human Factors Study Group of ACSNI*. Suffolk: HSE Books.

Argyris, C. (1982). *Reasoning, Learning, and Action: Individual and Organization*. San Francisco: Jossey-Bass.

Beazley, M. (1983). *Organization Theory*. London: Mitchell Beazley.

Beer, S. (1985). *Diagnosing the system for organizations*. Chichester: John Wiley and Sons.

Blau, P. (1957). 'Formal organization: dimensions of analysis'. *American Journal of Sociology*, **63**, 58–69.

Broadbent, D.E. (1989). *Advisory Committee on the Safety of Nuclear Installations: Study Group on Human Factors. First Report of Training and Related Matters*. London: Health and Safety Commission.

Bryans, P. and Cronin, T.P. (1983). *Organization Theory*. London: Mitchell Beazley.

Buchanan, D.A. (1979). *The Development of Job Design Theories and Techniques*. Farnborough: Saxon House.

Buchanan, D.A. and Huczynski, A.A. (1985). *Organizational Behaviour*. Engelwood Cliffs, NJ: Prentice-Hall.

Cohen, A., Smith, M. and Cohen, H.H. (1975). *Programme Practices in High Versus Low Accident Rate Companies – An Interim Report*. US Department of Health, Education and Welfare, Publication No 75–185 Cincinnati, OH: National Institute for Occupational Safety and Health.

Cox, T. (1978). *Stress*. London: Macmillan Press.

Cox, T. (1993). *Stress Research and Stress Management: Putting Theory to Work*. Sudbury, Suffolk: HSE Books.

Cox, T. and Griffiths, A. (1995). 'The nature and measurement of work stress: theory and practice'. In *The Evaluation of Human Work: A Practical Ergonomics Methodology* (J. Wilson and N. Corlett, eds), London: Taylor and Francis.

Cox, T. and Howarth, I. (1990). 'Organizational health, culture and helping'. *Work & Stress*, **4**.

Cox, T., Boot, N., Cox, S. and Harrison, S. (1988). 'Stress in schools: an organizational perspective'. *Work and Stress*, **4**, 353–362.

Cox, T., Cox, S. and Boot, N. (1989). 'Stress in schools: a problem solving approach'. In *Stress and Teaching* (M. Cole and S. Walker, eds), Milton Keynes: Open University Publications.

Cox, T. and Leiter, M. (1992). 'The health of healthcare organizations'. *Work and Stress*, **6**, 219–227.

Cox, T., Leather, P. and Cox, S. (1990). 'Stress, health and organizations'. *Occupational Health Review*, **23**, 13–18.

Cox, T., Kuk, G. and Leiter, M.P. (1992). 'Burnout, health, work stress, and organizational healthiness'. In *Professional Burnout: Recent Developments in Theory and Research* (W. Schaufeli and C. Maslach, eds), New York: Hemisphere.

De Michiei, J., Langton, J., Bullock, K. and Wiles, T. (1982). 'Factors associated with disabling injuries in underground coalmines'. *Mine Safety and Health Administration*, p. 72.

Duncan, W.J. (1981). *Organizational Behaviour*, 2nd edn. Boston, MA: Houghton Mifflin.

Emery, F.E. and Trist, E.L. (1960). 'Sociotechnical systems'. In *Management Science, Models and Techniques* (C.W. Churchman and M. Verhulst, eds), London: Pergamon Press.

Emery, F.E. and Trist, E.L. (1981). 'Sociotechnical systems'. In *Systems Behaviour*. (Open Systems Group, eds), London: Harper & Row.

Flood, R.L and Jackson, M.C. (1991). *Creative Problem Solving: Total Systems Intervention*. Chichester: John Wiley and Sons Ltd.

Hackman, J.R. (1986). 'The psychology of self-management in organizations'. In *Psychology and Work: Productivity, Change, and Employment* (M.S. Pallak and R. Perloff, eds), Washington, DC: American Psychological Association.

Health and Safety Executive (1981). *Managing Safety: A Review of the Role of Management in Occupational Health and Safety by the Accident Prevention Advisory Unit of H.M. Factory Inspectorate*. London: HMSO.

Health and Safety Executive (1991). *Successful Health and Safety Management*. HS(G)65, London: HMSO.

Hendry, C. (1991). 'Corporate strategy and training'. In *Training and Competitiveness* (J. Stevens and R. MacKay, eds) pp. 79–110, London: Kogan Page.

Herbst, P.G. (1954). 'The analysis of social flow systems'. *Human Relations*, **7**, 327–336.

Herbst, P.G. (1962). *Autonomous Group Functioning*. London: Tavistock Publications.

Hidden, A. (1989). *Investigation into the Clapham Junction Railway Accident*. London: HMSO.

Holland, J. (1985). 'Making vocational choices: a theory of careers'. In *Handbook of Life Stress, Cognition, and Health* (S. Fisher and J. Reason, eds), Chichester: John Wiley.

Howarth, C.I. (1995). 'Complexity theory and organizational health'. *International Forum for Organizational Health Newsletter*, August, 9–14.

Janis, I. and Mann, L. (1977). *Decision Making: A Psychological Analysis of Conflict, Choice, and Commitment*. New York: Macmillan.

Keenan, V., Kerr, W. and Sherman, W. (1951). 'Psychological Climate and Accidents in an Automotive Plant'. *Journal of Applied Psychology*, **35**, 108–111.

Lee, T.R. (1993). 'Psychological aspects of safety in the nuclear industry'. The Second Offshore Installation Management Conference 'Managing Offshore Safety', 29 April, Aberdeen.

Lee, T.R. (1995). 'The role of attitudes in the safety culture and how to change them'. Conference on 'Understanding Risk Perception', Offshore Management Centre, Robert Gordon University, 2 February, Aberdeen.

Lewin, K. (1951). *Social Theory and Social Structure*. New York: Harper.

March, J.G. and Simon, H.A. (1958). *Organizations*. New York: John Wiley.

Miller, E.J. and Rice, A.K. (1967). *Systems of Organization; The Control of Task and Sentient Boundaries*. London: Tavistock Publications.

Ministerial Statement on the Sizewell B nuclear power station (1987) *Atom*, **367**(36).

Open Systems Group (1981). *Systems Behaviour*. London: Harper and Row.

Perrow, C. (1967). *Organizational Analysis: A Sociological View*. London: Tavistock.

Peters, R.H. (1989). *Review of Recent Research on Organisational Factors and Behavioral Factors Associated with Mine Safety*. Bureau of Mines Inf. Circ., US Department of the Interior, Report IC9232.

Pidgeon, N.F. (1991). 'Safety culture and risk management in organizations'. *Journal of Cross-Cultural Psychology*, **22**, 129–140.

Porter, L.W., Lawler, E.E. and Hackman, J.R. (1975). *Behaviour in Organizations*. New York: McGraw-Hill.

Reason. J. (1995). 'A systems approach to organizational error'. *Ergonomics*, **38**(8), 1708–1721.

Rice, A.K. (1963). *The Enterprise and its Environment*. London: Tavistock Publications.

Rohner, R.P. (1984). 'Towards a conception of culture for cross-cultural psychology'. *Journal of Cross-Cultural Psychology*, **15**, 111–138.

Schein, E.H. (1985). *Organizational Culture and Leadership*. San Francisco: Jossey-Bass.

Scott, R.W. (1981). *Organizations: Rational, Natural and Open Systems*. Engelwood Cliffs, NJ: Prentice-Hall International.

Scott, W.G. and Mitchell, T.R. (1972). *Organization Theory*. Homewood, IL: Irwin-Dorsey.

Smith, M., Cohen, H., Cohen, A. and Cleveland, R.J. (1978). 'Characteristics of successful safety programmes'. *Journal of Safety Research*, **10**, 87–88.

Trist, E.L. and Bamforth, K.W. (1951). 'Some social and psychological consequences of the longwall method of coal-getting'. *Human Relations*, **4**, 3–38.

Trist, E.L. and Murray, H. (1948). *Work Organization at The Coal Face: A Comparative*

Study of Mining Systems. Doc. No. 506. London: Tavistock Institute of Human Relations.

Trist, E.L., Higgin, G.W., Murray, H. and Pollock, A.B. (1963). *Organizational Choice*. London: Tavistock Publications.

Turner, B.A., Pidgeon, N.F., Blockley, D.I. and Toft, B. (1989). 'Safety culture: its position in future risk management'. Paper to: Second World Bank Workshop on Safety Control and Risk Management, Karlstad, Sweden.

Waldrop, M.M. (1989). *Complexity*. New York: Simon and Schuster.

Waring, A.E. (1989). *Systems Methods for Managers: A Practical Guide*. Oxford: Blackwell.

Woodward, J. (1965). *Industrial Organization: Theory and Practice*. London: Oxford University Press.

Zohar, D. (1980). 'Safety climate in industrial organisations – theoretical implications'. *Journal of Applied Psychology*, **65**(1), 96–102.

6
Jobs and tasks

Introduction

It has been argued that the psychology and management of safety can be best understood within a framework based in systems theory which describes work systems in terms of the person in their job in their organization. The previous chapter discussed issues related to the organization and the subsequent chapter focuses on the person. This chapter considers the job from a task system perspective (Ilgen and Hollenbeck, 1991).

The task system perspective views organizations in terms of the objectives to be achieved and the tasks involved in achieving them. Those tasks are grouped or clustered into jobs. Jobs do not always represent particular objectives; indeed, a job may involve tasks derived from a number of different objectives. In most organizations the basic unit of analysis and management is the job. Tasks are part of the software of organizations and, as such, part of the technical system.

It should be obvious from the task systems perspective that in order to accomplish any organizational objective, there must be coordination and control over the various jobs which contribute to the achievement of that objective through their constituent tasks. Thus attention must be paid not only to individual jobs and tasks but also to the relationship between jobs and the moderation of that relationship. The latter is a part of the control function of management.

This chapter focuses on the nature of the jobs that people do in organizations in terms of the tasks which make up those jobs. In doing so, it considers the ways in which different tasks combine to form jobs, and how such jobs might be analysed to reveal their demand characteristics in terms of their psychological and social requirements. These psychosocial characteristics of jobs are considered again in Chapter 8. Research on such job characteristics has been largely motivated by concern for the design and management of work which is satisfying and without threat to wellbeing. However, this discussion can be directed into a more general consideration of the nature of safe work in a way which allows lessons to be drawn out for the management of safety at work. Finally, considerations of human reliability assessment complete the initial discussions of jobs and tasks. This is further explored in later chapters.

Job descriptions and taxonomies

The study of jobs has a long history, and has been dominated by industrial engineers, ergonomists and occupational psychologists often primarily interested in developing methods of job analysis and description for the purpose of building taxonomies. Analysing, describing and understanding the attitudes, knowledge, skills and other behaviour required by any job is a prerequisite for recruitment, selection and training, job and equipment design, strategic management and the management of safety at work (see Table 6.1; also Ilgen and Hollenbeck, 1991). Job descriptions and taxonomies have thus been treated as critical tools for the effective discharge of these organizational functions. Like all taxonomic systems, those developed for analysing and describing jobs are arbitrary with their value depending on the extent to which they are useful for the purposes for which they have been developed and to which they are applied.

No one universal taxonomy exists. Fleishman's seminal work in this area (see, for summary, Fleishman and Quaintance, 1984) began as an attempt to

Table 6.1 Some of the uses of information obtained through job analysis

Personnel Procedures	Work equipment and procedures	Strategic management	Other uses
Personnel recruitment, selection and placement	Engineering design	Organizational planning	Planning educational curricula
	Methods design	Manpower planning and control	Vocational counselling
Training and personnel development	Job design		
		Safety management	Job classification systems
Performance measurement and appraisal			Risk assessment

develop an overarching taxonomy of tasks that would provide the framework for classifying all tasks. However, his attempt to provide a universal system led to the contrary conclusion that several different systems are needed to capture the important and various dimensions of work tasks (Fleishman and Quaintance, 1984). Similarly, it has also been concluded that several different taxonomies are required to adequately capture the diversity of jobs which exist within organizations.

Inventories derived from job and task taxonomies proved popular in the past and typically consisted of a list of tasks that were deemed pertinent to a particular type of job. The completion of the inventory involved rating each task against several factors, such as the *frequency* with which a task was performed, the *time* spent on the task, its importance (etc.). Task inventories were, for example, extensively used by military establishments (see, for example, Marsh, 1962 for the description of task inventories used by the US Airforce). Both job and task inventories, and the taxonomies on which they were based, usually depended on job analysis for their content.

Job analysis

Techniques for job analysis have been discussed in some detail by Kirwan and Ainsworth (1992) and are briefly outlined below. Kirwan and Ainsworth (1992) distinguish between task data collection techniques, such as activity sampling and critical incident technique, task description techniques, such as hierarchical and operational sequence diagrams, and task simulation methods, such as computer modelling. Central to all of these is the question of measurement; suffice here to point out the importance of only using measures which have been properly developed, and which have been proved reliable, valid and unbiased (fair).

Job analyses deconstruct jobs into tasks, and tasks into subtasks, and subtasks into activities. This process of deconstruction is usually stopped when the integrity and meaning of a set of activities is challenged. If not, the analysis would become more and more detailed at lower and lower levels until the whole process lost meaning in terms of the understanding and management of work: activities into movements, movements into muscle contractions, and so on. The level of analysis and the stop point are not fixed features of such analyses, but are determined in part by the objectives of the analysis which may only require, for example, a gross distinction between major job components. This point may be made more clearly by comparing the typical job description which has a job analysis for the design of a selection procedure with that necessary for a time and motion study.

Such systems of analysis and description should be capable of handling not only objective data on jobs, but also subjective data: individuals' perceptions and descriptions of their jobs may be captured through the use of knowledge elicitation techniques.

Task description techniques are techniques which structure and present the information collected through job analyses in a systematic format. It is not uncommon to see such analysis data presented in diagrammatic form, including: charting and network diagrams, hierarchical and operational sequence diagrams, link and timeline diagrams. Often jobs and tasks are described using some combination of top-down tree diagrams (hierarchical method) and flow-charts (operational sequence method) as appropriate (see Chapter 11). Jobs which have more complex structures are usually summarized using top-down tree diagrams, while those that have more predictable structures and repetitive sequences of tasks and activities are usually presented in flow-chart form. These are the extremes, and most jobs can be presented as combinations of both methods: the top-down tree structures summarizing the more important job components or tasks, and the flow-charts exploring the more predictable sequences within any or all of those tasks.

The pictorial representation of job analysis data offers an easy form of model for health and safety purposes (see Chapter 11). The limitations of such models are that they are fixed, often two dimensional, and cannot be interrogated and explored. However, it is increasingly possible to take the sort of data described above and develop relatively simple 'expert systems' to describe jobs in a more adequate way. Such systems should, in principle, offer a more dynamic and interactive model of the job, and one in which the relationships between tasks (subtasks, etc.) can be explored. The nature and application of expert systems has been described by Waterman (1986).

Job analysis may be the precursor to risk assessment (see Chapter 2) being used to identify jobs and tasks for subsequent analysis. The level of analysis required (see above) may be determined by the nature of the situation under consideration. For example, a safety audit of an office environment may only require a relatively high level of analysis specifying the different jobs involved, their defining tasks, and any 'risky' tasks. However, investigation of machine-related injuries in a repetitive assembly task may require a detailed analysis of movement patterns for each task in each job involving that machine. Information obtained by job analysis can also serve a variety of other purposes which relate indirectly to safety (see Table 6.1). For example, effective recruitment selection and job placement is an important mechanism for ensuring safety.

Objective characteristics of jobs

In addition to the description of the actual tasks involved, jobs are commonly described, both in everyday conversation and more scientifically, in terms of several major dimensions, *inter alia*, according to:

1 the level of skill and knowledge demanded by the tasks involved (see, for example, Rasmussen's (1980) SRK model in Chapter 7);
2 the socioeconomic status implied by the job; and
3 the level of task complexity involved.

For example, contrasts are often made between (1) unskilled, semi-skilled and skilled work, between (2) manual, non-manual, managerial and professional work, and between (3) simple and complex work, and repetitive and varied work. Although these three dimensions provide much of the common language of jobs, there are many other possible dimensions of description. The three under consideration here are objective characteristics of jobs.

Not only do jobs vary in terms of each of these objective characteristics but also in the ways in which these characteristics combine. Although the different characteristics could be independent, in practice they tend to be related in the way in which they combine. Perhaps the most obvious example is provided by the contrast between, say, repetitive assembly jobs and those of the managers who are responsible for such assembly workers. Traditional repetitive assembly jobs involve the fixed (and predictable) combination of sequences of unskilled or semi-skilled manual tasks. Often much of these jobs is made up of a single task repeated with a short work cycle, possibly of thirty seconds or less (T Cox, 1985). The associated managerial jobs probably involve a variety of more or less complex tasks combined in a largely unpredictable way with little evidence of repetition, and with very long cycles where any repetition can be detected. Such cycles may be, at the least, several hours, if not days or weeks, long. In this way, the dimensions mentioned above may, in practice, overlap and form clusters of typical jobs. For example, one such cluster might be: unskilled or semi-skilled, manual, simple and repetitive, low socioeconomic status. Another might be: skilled, managerial, complex and varied, relatively high socio-economic status. A third, not mentioned above, might be: requiring high level of both knowledge and skill, professional, complex and varied, very high socioeconomic status. Thus a possible taxonomy of jobs might be built up using these objective characteristics (see Table 6.2).

Table 6.2 Taxonomy by objective job characteristics

Including:	
1	Level of knowledge and skill required
2	Associated occupational and socioeconomic status
3	Variety of tasks involved
4	Complexity of job

If such a taxonomy is to be proved valuable, then two questions have to be answered: first, how can the different characteristics involved be reliably and validly measured, and, second, can the utility of the taxonomy be proved through use? The first of these questions is considered here.

Measurement of objective characteristics

The level of knowledge and skill required by a task can be assessed in terms of two criteria: first, what is the minimal level of education required for reasonable performance of the task (no secondary schooling, CSEs, GCSEs, A levels, BTEC, first degree or higher degree), and, second, does the task require any special training or education, and of what nature and duration (no or yes, manual or intellectual skills, short, medium or long duration)? Thus a short cycle assembly task might be assessed in terms of: no secondary schooling or special training required. The junior or middle manager's typical task might require: A levels and special education of medium duration. The surgeon's typical task might require, at least, a first degree and special manual and intellectual training of long duration.

Occupational and related socioeconomic status have been conveniently described by the Registrar General's occupational taxonomy (1970) in terms of six defined categories: unskilled manual, semi-skilled manual, skilled manual, skilled non-manual, managerial and professional. This has been developed to include a further category, junior non-manual (OPCS, 1995). Most recognized jobs (in the UK) have been fitted into this scheme, and together the categories and related jobs are described, for example, in the *Occupational Mortality Statistics Decennial Supplement 1970–1972* (OPCS, 1978). Thus the assembly task might be categorized as unskilled or semi-skilled manual, the middle manager's job as managerial, and the surgeons job as professional.

Issues of variety of tasks and complexity of jobs can be operationalized in terms of the number of different tasks involved in any job, the probability of any one task following any other task and the rules or contingencies which determine those probabilities. Variety may be relatively simply operationalized in terms of the number of different tasks involved, and complexity in terms of the associated probabilities, rules or contingencies. The greater the number of rules and the lower the probabilities involved, arguably, the greater the complexity. There is an important question here as to how variety and complexity so measured objectively relate to their perception and experience by the individual. This issue is dealt with below. The repetitive assembly job might involve very few different tasks with high probabilities linking the different tasks (into an obvious sequence) and with only a few simple rules determining those linkages. The middle managers job, and that of the surgeon, might, by contrast, involve a large number of different tasks with low probabilities linking tasks and with hierarchies of contingencies determining those linkages. This sort of analysis can be useful in building computer-based models of jobs.

Combining the objective dimensions presented in Table 6.2 and discussed above can provide a description of any particular job and allow different jobs to be grouped into job types. Part of the challenge of safety management is to explore the nature of the jobs involved in the work system, and a variety of analysis techniques are available for this purpose; these are considered in Chapter 11. However, the objective dimensions of jobs and tasks are not the sole determinants of work behaviour and safety. Attention also has to be paid to workers' perceptions of their jobs and tasks.

Perceived job characteristics

Perceived job characteristics have been studied generally using one of three methodologies. The first, and perhaps the least adequate because of its inherent lack of quantification, relies on individuals' qualitative accounts, verbal or written, of their jobs elicited usually through semi-structured or structured interviews or by questionnaires or diaries. The main job characteristics are then 'drawn out of' these accounts. Whatever its shortcomings, the strength of this approach lies in the lack of constraint imposed on those accounts by the analyst. The accounts represent the job as seen through the individual's eyes and not the analyst's. The second method seeks the person's endorsement (or other response) to a list of job characteristics imposed on them by the analyst. This list might be constructed through review of the professional and scientific literatures relevant to the job under consideration, expert opinion, or as a result of a particular theoretical position or working model. The strength of this approach is that it offers a standard format for enquiry and, usually, quantified data because of the nature of the response scales used. Its major shortcoming is that it imposes a particular view of the job on the individual: it runs the risk of seeing the job through the analyst's eyes and not the person's. The third approach is essentially the sensible compromise between the former two. It is a two stage enquiry. The first stage seeks to identify through a variety of methods, interview, diaries, questionnaires, the characteristics of the job under consideration, and, in the second stage, these are presented in the form of a structured instrument to those carrying out the jobs and with a response scale appropriate for the capture of quantified data.

An example of the use of this latter methodology is provided by the second author's development of a Job Description Checklist (JDCL) for repetitive work (see T Cox, 1985; T Cox and Mackay, 1979). In a major study of the psychological and physiological effects of short cycle repetitive work, an adjective checklist was developed for the assessment of workers' descriptions of their jobs. This was developed in three stages. In the first stage, discussions were held with groups of workers involved in short cycle repetitive work during which the adjectives they used to describe their jobs were noted. These were then used to construct an adjective checklist with an associated intensity-based response scale. The list was then discussed with other groups of workers, and piloted and modified to take into account the feedback made available. In the second stage, the modified checklist was distributed to a large sample of workers drawn from populations engaged on a variety of types of short cycle repetitive work. The data from this sample was then subjected to exploratory factor analysis (see Ferguson and T Cox, 1993). Two factors emerged from this analysis: job pleasantness and job difficulty. The internal reliability and the test–retest reliability of these two factors and associated scales were then established. In the third stage, data collected using the JDCL were broken down by age, gender and type of job, and also related to workers' psychological and physiological reactions to those jobs. These subsequent analyses suggested that the simple model (see Figure 6.1) underpinning the JDCL was not only reliable but also had some validity and utility. The model's simplicity was important in that it reflected the way in which workers perceived and described their jobs, and the fact that such descriptions could be related not only to other aspects of

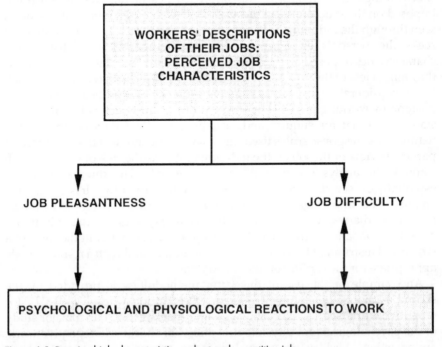

Figure 6.1 Perceived job characteristics – short cycle repetitive jobs

their psychological reactions but also to their physiological reactions suggests that the methodology used was appropriate.

Data from studies such as that described above have informed the development of particular models of the psychological and social dimensions of jobs. Two particular models are described here: the job characteristics and the person–environment fit models.

Job Characteristics Model

Perhaps one of the earliest and most influential models of job characteristics is that of Hackman and Oldham (1976, 1980).

The Job Characteristics Model (JCM) has its origins in a study by Turner and Lawrence (1965) that examined the relationship between attributes of tasks and employees' reactions to their work. They constructed measures of six task attributes that they predicted to be positively related to employee satisfaction and attendance. With a sample of forty-seven jobs, they empirically derived a summary index – the requisite task attribute index – which they then used to predict the two outcome measures. The index was the linear combination of the individual task attribute scores. Hackman and Lawler (1971) followed up Turner and Lawrence's (1965) work, narrowing the set of six requisite task attributes to four. They proposed that employees who desired growth and involvement in their work would respond more favourably toward jobs that were high on their four requisite task attributes than those that were low on these attributes. Hackman and Oldham (1976, 1980) further extended and refined the JCM. It is their version of the model which has largely caught the popular imagination, and it is that version which is described here. Their basic model is presented in Figure 6.2.

At the most general level, five job characteristics are seen as prompting three psychological states which, in turn, lead to a number of personal and work-relevant outcomes. The links between these variables are moderated by individual differences.

The three psychological states – experienced meaningfulness of work, experienced responsibility for the outcomes of work, and knowledge of results – are the core of this model. The presence of these states creates the motivation to persevere with work until such time as those states no longer exist. The job characteristics which foster the emergence of these key psychological states are: skill variety, task identity and task significance (meaningfulness of work), autonomy (responsibility for outcomes), and feedback from job (knowledge of results).

A general criticism of the job characteristics approach to work design has been that it appears to have encouraged a relatively narrow theoretical perspective which has largely ignored contextual factors (Roberts and Glick, 1981; Wall and Martin, 1994).

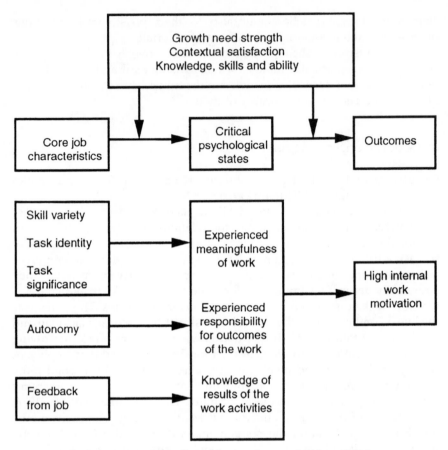

Figure 6.2 Job characteristics model (adapted from Hackman and Oldham, 1976)

An important exception is the linkage between the job characteristics approach and interest in autonomous group work, and sociotechnical theory (see Chapter 5). Both Rousseau (1977) and Cummings (1982) called for a synthesis of these two approaches. Indeed, Hackman (1977) has suggested that autonomous work groups are likely to prove more powerful than individual forms of job design since they can encompass much larger and more complete pieces of work. He subsequently extended the JCM to apply at the group level (Hackman, 1983).

Person–Environment Fit (P–EF)

The Person–Environment Fit (P–EF) model of French *et al.* (1987), and their various colleagues in Michigan and elsewhere, directly addresses, among other things, the relationship between objective and perceived (subjective) job characteristics.

The P–EF model makes two basic distinctions in relation to our understanding of work; the first is between the characteristics of the person and those of the (work) environment, and the second is between objective characteristics (those that can be determined and measured objectively) and subjective characteristics (those that are reflected in the person's perceptions and which are measured as such). A clear distinction is made in this theory between objective reality and subjective perceptions, and between environmental variables (E) and person variables (P).

The model (see Figure 6.3) recognizes a dynamic relationship – an ongoing interaction – between the person and their work environment, and assesses the goodness of that interaction in terms of the degree of fit (or match)

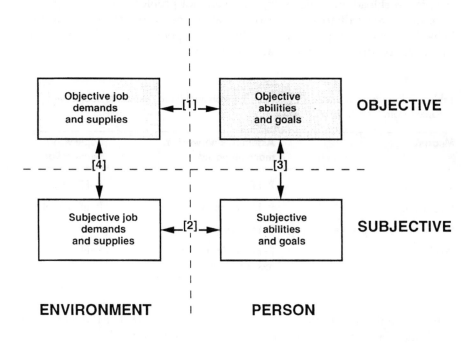

Figure 6.3 Person–Environment Fit (P–EF) Model (adapted from French *et al.*, 1982)

between the person and the (work) environment. Lack of fit can occur in four different ways, and each may challenge the workers' safety and health. There can be both a lack of subjective and objective P–EF. There can also be a lack of fit between the objective environment (reality) and the subjective environment (hence lack of contact with reality) and also a lack of fit between the objective and subjective persons (hence poor self-assessment).

Table 6.3 presents data drawn from French *et al.* (1982) which describes job differences in relation to the degree of subjective fit for four measures: job complexity, role ambiguity, responsibility for people and workload. It can be seen from Table 6.3 that assembly line workers and white collar supervisors (managers) differ markedly in relation to job complexity and responsibility for people with the former having jobs which are perceived to be too simple and with too little responsibility for people. The latter, by contrast, have jobs which are perceived to be too complex and have too much responsibility for people. Workload and role ambiguity were not problems for these groups. Poor fits were predictive of reactions to work and work- and health-related behaviour. For example, overall, jobs with a poor fit for job complexity predicted job dissatisfaction relatively strongly.

Table 6.3 Results from analysis of Person–Environment Fit in relation to subjective fit (adapted from French, *et al.*, 1982)

Measure	Assembly-line worker: machine paced	Supervisor: white collar
Job complexity	P–EF very poor	P–EF poor
Role ambiguity	P–EF acceptable	P–EF acceptable
Responsibility for people	P–EF poor (too little)	P–EF poor
Workload	P–EF OK (at time of measurement)	P–EF poor

Interestingly, the empirical evidence tends to suggest that the subjective P–EF may be a stronger determinant of behaviour and subsequent health and safety than the objective P–EF (French *et al.*, 1987). Within this model, the environment is characterized in terms of the demands it places on the person, and what it supplies to meet the person's needs and expectations. Much attention has been paid to the nature of these demands, and this issue is discussed in more detail in Chapter 8 on work-related stress. Also important in this respect is the control that workers have over their work, and the support that they receive in coping with work. These are also discussed in Chapter 8.

French *et al.* (1982) have reported on a large US survey of work stress and health in twenty-three different occupations and a sample of 2010 male workers. The survey was framed by the P–EF theory and, in their summary, they commented on a number of questions of both theoretical and practical importance. In particular, they argued that their subjective measures reflected objective realities and that they mediated the effects of (objective) work on health. Their data showed that there was a good correspondence between the objective measures of work (hazards) and the subjective measures of stress and that the effects of those objective measures on self-reported health could be very largely accounted for by the subjective measures. Objective measures only accounted for some 2–6 per cent of the variance in self-reported health beyond that accounted for by the subjective measures.

Models such as those briefly described above both inform and frame our use of job analysis techniques to build models of particular jobs. These models can be used as a basis for identifying both hazards and opportunities for human error. Such information is important for the management of safety, but insufficient on its own for the assessment of risk. What is required in addition to, and following from, the job analysis, is the quantification of

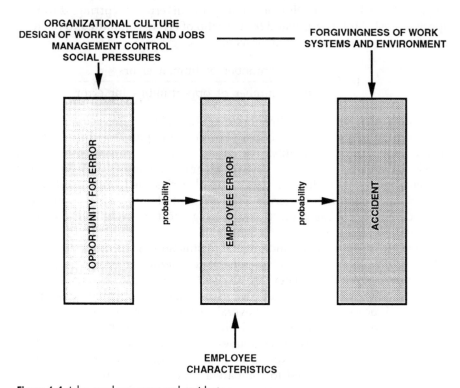

Figure 6.4 Jobs, employee error and accidents

the reliability of the employee in carrying out particular tasks, or generally completing the job. The questions are how likely is it that the employee makes an error given the opportunity for error, and how likely is it that an employee error will lead to an accident or other challenge to safety? These issues are represented diagrammatically in Figure 6.4, and discussed in the following sections in relation to the techniques used to answer the two questions posed above. They are also considered in the next chapter in relation to individual cognition and behaviour.

Measurement of human error probability

Attempts to quantify human reliability have been incorporated into systems thinking since the late fifties and originated in the aerospace industry. The majority of the work has taken place within those industries which are perceived as 'high risk', for example, aerospace, chemical and nuclear process industries. It has been related to the probabilities of human error for critical functions and particularly in emergency situations (Humphreys, 1988). The most common measure of human reliability is the Human Error Probability (HEP). It is the probability of an error occurring during a specified task. HEP is estimated from a ratio of errors committed to the total number of opportunities for error as follows:

$$ HEP = \frac{\text{number of human errors}}{\text{total number of opportunities for error}} $$

The Safety and Reliability Directorate's (SRD) *Human Reliability Assessors Guide* (Humphreys, 1988) describes several techniques (see Table 6.4) for determining human reliability. The guide is written at a user level and provides detailed case studies for each technique, together with additional reference material. Hollnagel (1993) has also classified human reliability assessment techniques in terms of their central approaches into those methods which are based on (i) engineering and (ii) psychological models (see Table 6.4).

However, despite this fundamental distinction (Hollnagel, 1993), the majority of human reliability assessment techniques are based *in part* on behavioural psychology. They are derived from empirical models using statistical inference but are not always adequately validated (Center for Chemical Process Safety, 1989). Several of the models are also based on expert judgements (see Table 6.4) which may be subject to bias. Bias may be overcome by applications of techniques such as paired comparisons. This method does not require experts to make any quantitative assessments, rather the experts are asked to compare a set of pairs for which HEPs are required and for each pair must decide which has the higher likelihood of

Table 6.4 Techniques for assessing human reliability

(i)   **Methods based on engineering models**

a   **Expert Judgement Models**
*Absolute Probability Judgement (APJ)
*Paired Comparison (PC)
*Influence Diagram Approach (IDA)
*Success Likelihood Index Method (SLIM)
Sociotechnical Approach to Assessing Human Reliability (STAHR)

b   **Analytical Methods**
*Technique for Human Error Rate Prediction (THERP)
Accident Sequence Evaluation Programme (ASEP)
Task Analysis-Linked Evaluation Technique (TALENT)

(ii)   **Methods based on psychological/cognitive models**
System Respond Analyses/Generic Error Modelling System (SA/GEMS)
Systematic Human Error Reduction and Prediction Approach (SHERPA)

---

* Described in Humphreys (1988)

error (Humphreys, 1988). However, providing that the limitations of HEPs are recognized, it is often better to have carried through the process of deriving an 'acceptable' figure than to dismiss the task as impossible. Although it is important to note that such measurements should be applied with caution and in a way which takes account not only of the complexity of the overall system and of the potential accident process, but also of the exact nature of the task to which it refers. Rosness *et al.* (1992) have described the elements of an ideal method for human reliability analysis (and assessment) as follows:

1 The method should use the individual task-steps of a formal operating procedure as a starting point and consider the error modes which can arise at each step. These steps can be determined using task analysis techniques.
2 If the consequences of error are significant but prevention measures are expensive then the method should provide an in-depth search for possible causes (fault-tree analysis, as described in Chapter 3, can be used to support this requirement).
3 The method should take into account particular patterns of multiple actions which arise from knowledge intensive tasks (see earlier).
4 Interaction effects should be considered, for example, seemingly simple actions (unsafe act tokens) may interact with latent failures (Reason, 1990) in the system to trigger serious accidents.

5 General weaknesses in the person-machine interface (see Chapter 11) should be identified.
6 The method should incorporate much of the current knowledge on human error and cognition (see next chapter).
7 The method should give explicit recommendations on the prediction of Human Error Probabilities (HEP).

One of the most commonly used techniques for determining HEPs, THERP (Swain and Guttman, 1983), was developed mainly to assist in determining human reliability in nuclear power plants, and addresses many of these requirements. It is described below.

Technique for human error rate prediction (THERP)

This technique, developed by Swain and Guttman (1983) can be broken down into a number of discrete stages which require the analyst to provide:

1 a system's description, including goal definition and functions;
2 a job and task analysis by personnel to identify likely error situations;
3 an estimation of the likelihood of each potential error, as well as the likelihood of its being undetected (taking account of performance shaping factors, see Table 6.5 and Chapter 8);
4 an estimate of the consequences of any undetected error; and
5 suggested and evaluated changes to the system in order to increase success probability.

Figure 6.5 provides a flow chart for THERP.
    The original definition of the system's goals (see Chapter 4) is followed by a job and task analysis. In the task analysis procedures for operating and maintaining the system are partitioned into individual tasks. Other relevant information, for example equipment acted upon, action required of personnel, and the limits of operator performance, is documented at this stage. The task analysis is followed by the error identification process. The errors likely to be made in each task step are identified and non-significant errors (those with no important system consequences) are ignored.
    The third stage is the development of an event tree: each likely error is entered sequentially as the right limb in the binary branch of the event tree (see Figure 6.6).
    The first potential error starts from the highest point of the tree at the top of the page. Each stage of the left limb thus represents the probability of success in the task step and each right limb represents its failure probability. To determine the probability of the task being performed without error, a complete success path through the event tree is followed.

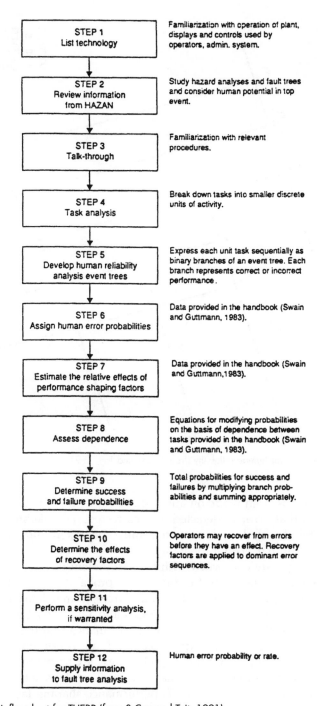

Figure 6.5 A flowchart for THERP (from S Cox and Tait, 1991)

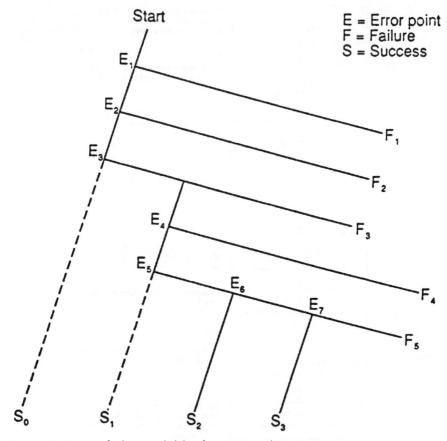

Figure 6.6 Event tree for human reliability (from S Cox and Tait, 1991)

Once an error has been made on any task, the system is presumed to have failed, unless that error is detected and corrected. The likelihood that an error will be detected and corrected must be taken into account by modifying the initial error probability.

The final stages are concerned with the assignment of error probabilities (Swain, 1987). Here the analyst estimates the probability of occurrence for each error, making use of all available data sources, formal data banks, expert judgements, etc. (for further details see S Cox and Tait, 1991). Such estimates take into account the relative effects of performance-shaping factors: (for example, stress, proficiency, experience) and all conditions which are assumed to affect task performance significantly (see Chapter 8). This process is often fairly arbitrary and may be one of the weakest steps in the procedure. According to Miller and Swain (1987) performance-shaping factors include any which have the potential to influence human performance. Such factors may be either internal (i.e. relate to the individual) or external, relating to the environment (see Table 6.5). External factors (Miller and Swain, 1987) are considered to have the greater impact on human reliability (see Figure 6.4). Many of these factors are considered in later chapters.

They also make an assessment of task dependence. Except for the first branch of the event tree, all branches represent conditional probabilities, with task/event interdependence directly affecting success/failure probabilities. Thus, each task must be analysed to determine its degree of dependency. Each end point of an event tree is labelled as a task success or failure, qualified probabilistically, and combined with other task probabilities to formulate total system success/failure probabilities. Details of mathematical calculations can be found in the *Human Reliability Assessors Guide* (Humphreys, 1988) or in the THERP *Handbook* (Swain and Guttman, 1983).

Table 6.5 Performance shaping factors (PSFs) (after Miller and Swain, 1987)

Internal PSFs

Emotional state	Skill level/previous job history
Intelligence	Social factors
Motivation/attitude	Strength/endurance
Perceptual abilities	Stress level
Physical condition (health)	Task knowledge
Sex differences/age	Training/experience

External PSFs
Inadequate task design
Inadequate workspace and layout
Poor environmental conditions
Inadequate training and job aids
Poor supervision
Unrealistic deadlines

Although, because of its mathematical basis, this method of error rate prediction implies accuracy, its utility is only as reliable and valid as the reliability and validity of its various measures. If these are not meaningful or themselves accurate then there is the real possibility that the overall process will itself be 'meaningless'.

There have been various criticisms on the human reliability assessment approach. Meister (1985) has cautioned against the illusory level of precision that can be assigned to HEP measures given the variability of human behaviour. Other human factors specialists (see, for example, Sanders and McCormick, 1987 and Hollnagel, 1993) have pointed out its limitations. It is, despite being based upon expert judgement, a subjective approach and should be used sensibly alongside other system data for guiding decision making on risk.

Summary

This chapter has explored the nature of work tasks and jobs in terms of their objective and subject characteristics with some reference to the effects of these characteristics on the job incumbent. This task-focused perspective sees tasks as deriving from the organization's objectives, and grouping and clustering to form jobs.

Methods of job and task analysis and description are briefly reviewed, and the value of such techniques explored in relation to risk assessment and, also, the measurement of human error probability.

The job, and its constituent tasks, are the real link between the organization and the person in relation to its work systems.

References

Center for Chemical Process Safety (1989). *Guidelines for Chemical Process Quantitative Risk Analysis*. The American Institute of Chemical Engineers.

Cox, S. and Tait, N.R.S. (1991). *Reliability, Safety and Risk Management: An integrated approach*. London: Butterworth Heinemann.

Cox, T. (1985). 'Repetitive work: occupational stress and health'. In *Job Stress and Blue Collar Work* (C.L. Cooper and M.J. Smith, eds), Chichester: Wiley and Sons.

Cox, T. (1987). 'Stress, coping and problem solving'. *Work and Stress*, 1, 5–14.

Cox, T. and Mackay, C.J. (1979). 'The impact of repetitive work'. In *Satisfactions in Job Design* (R. Sell and P. Shipley, eds), London: Taylor and Francis.

Cummings, L.L. (1982). 'Organizational behaviour'. *Annual Review of Psychology*, 33.

Ferguson, E. and Cox, T. (1993). 'Exploratory factor analysis: a user's guide'. *International Journal of Selection and Assessment*, 1(2), 84–94.

Fleishman, E.A., and Quaintance, M.K. (1984). *Taxonomies of Human Performance*. New York: Academic Press.

French, J.R.P., Caplan, R.D. and van Harrison, R. (1982). *The Mechanisms of Job Stress and Strain*. New York: Wiley and Sons.

French, J.R.P., Caplan, R.D. and van Harrison, R. (1987). *Person–Environment Fit*. Wiley & Sons, Chichester.

Hackman, J.R. (1977). 'Work design'. In *Improving Life at Work* (J.R. Hackman and J.L. Shuttle, eds), Santa Monica, CA: Goodyear.

Hackman, J.R. (1983). 'The design of work teams'. In *Handbook of Organizational Behaviour* (J. Lorsch, ed), Englewood Cliffs, NJ: Prentice-Hall.

Hackman, J.R and Lawler, E.E. (1971). 'Employee reactions to job characteristics'. *Journal of Applied Psychology*, 55, 259–286.

Hackman, J.R., and Oldham, G.R. (1976). 'Motivation through the design of work: test of a theory'. *Organizational Behaviour and Human Performance*, **16**, 250–279.

Hackman, J.R., and Oldham, G.R. (1980). *Work Redesign*. Reading, MA: Addison-Wesley.

Hollnagel, E. (1993). *Human Reliability Analysis Context and Control*. London: Academic Press.

Humphreys, P. (1988). *Human Reliability Assessor's Guide*. Culcheth, Warrington: UKAEA.

Ilgen, D.R. and Hollenbeck, J.R. (1991). 'The structure of work: job design and roles'. In *Handbook of Industrial and Organizational Psychology* (M.D. Dunnette and L.M. Hough, eds), Palo Alto, CA: Consulting Psychologists Press.

Kirwan, B. and Ainsworth, L.K. (1992). *A Guide to Task Analysis*. London: Taylor and Francis.

Marsh, J.E. (1962). 'Job analysis in the United States Air Force'. *Personnel Psychology*, **37**, 7–17.

Meister, D. (1985). *Behavioural Analysis and Measurement Methods*. New York: Wiley.

Miller, D.P. and Swain, A.D. (1987). 'Human error and reliability'. In *Handbook of Human Factors* (G. Salvendy, ed) pp. 219–250, New York: Wiley.

Office of Population Censuses and Surveys (OPCS) (1978). *Occupational Mortality Decennial Supplement 1970–1972*. London: HMSO.

Office of Population Censuses and Surveys (1995). *Occupational Health Decennial Supplement* (F. Drever, ed.). London: HMSO.

Rasmussen, J. (1980). 'What can be learned from human error reports?' In *Changes in Working Life* (K.D. Duncan, M. Gruneberg and D. Wallis, eds), London: Wiley.

Reason, J. (1990). *Human Error*. Cambridge: Cambridge University Press.

Roberts, K.H. and Glick, W. (1981). 'The job characteristics approach to job design: a critical review'. *Journal of Applied Psychology*, **66**, 193–217.

Rosness, R., Hollnagel, E., Sten, T. and Taylor, J.R. (1992). *Human Reliability Assessment Methodology for the European Space Agency (STF75 F92020)*. Trondheim, Norway: SINTEF.

Rousseau, D.M. (1977). Technological difference in job characteristics, employee satisfaction motivation: a synthesis of job design research and sociotechnical systems theory. *Organizational Behaviour and Human Performance*, **19**, 18–42.

Sanders, M.S. and McCormick, E.J. (1987). *Human Factors in Engineering Design*, 6th edn. New York: McGraw Hill.

Swain, A.D. (1987). *Accident Sequence Evaluation Procedures. Human Reliability Analysis Procedure*. Washington, DC: Sandia National Laboratories, NUREG/CR-4722, US Nuclear Regulatory Commission.

Swain, A.D. and Guttman, M.E. (1983). *Handbook of Human Reliability Analysis with Emphasis on Nuclear Power Plant Applications*. Washington, DC: NUREG/CR-1275 US Nuclear Regulatory Commission, p. 222.

Turner, A.N. and Lawrence, P.R. (1965). *Individual Jobs and the Worker*. Cambridge, MA: Harvard University Press.

Wall, T.D. and Martin, R. (1994). 'Job and work design'. In *Key Reviews in Managerial Psychology* (C.L. Cooper and I.T. Robertson, eds), Chichester: John Wiley and Sons.

Waterman, D.A. (1986). *Expert Systems*. Reading, MA: Addison-Wesley Publishing Company.

7
The person

Introduction

Chapter 7 concerns the role of the person in the authors' model of safety management and discusses the cognitive and emotional processes which together determine the person's behaviour. In doing so it considers the nature and role of attitudes in relation to that behaviour. Besides providing a basic understanding of cognition, emotion and behaviour, this chapter also addresses two issues of particular relevance to safety: human error (including violations) and risk perception. What is being discussed is the psychology of the person, as opposed to that of the job or the organization, and this chapter begins by presenting a model of the person which can be used to support our understanding and management of safety at the individual level.

The person as an information processor

Best (1992) has argued that many psychologists treat the person as if they were an 'information processing system' (see, for example, Simon, 1969). Such an approach, which can be contrasted with that of the more recent connectionist school (see later), is presented schematically in Figure 7.1. Figure 7.1 is essentially an information flow diagram. The arrows represent the flow of information through the system and the labelled boxes represent functional elements in the information processing chain. Inputs into the system are the various sources of external information and outputs from the system are actions (behaviours). Figure 7.1 represents the basic Human Information Processing (HIP) model which has been elaborated by authors such as Dodd and White (1980).

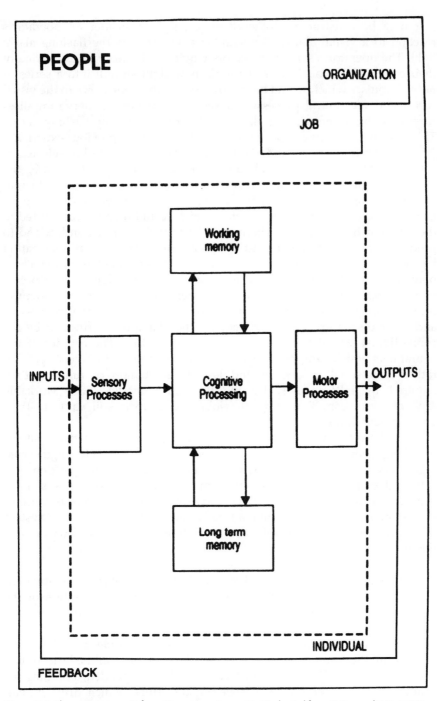

Figure 7.1 The person as an 'information processing system' (adapted from S Cox and Tait, 1991)

It may be constructive to work an example. Consider an operator's reaction to a visual alarm. The significant stimulus is the flashing alarm light. The operator's sensory receptors would be, in this case, the retina of their eyes which would convert the flashing light stimuli into a series of nerve impulses which would be transmitted via the optic nerve to the visual cortex and the central processing system. The pattern of nerve impulses carries the basic information about the light stimulus: the flashing alarm. This information would be interpreted in the context of previous experience (by reference to memory) and a response of some kind would be selected or constructed from the available repertoire of actions. In this case, the response might be a decision to push down on a button, with the right hand, to stop the alarm flashing. This decision on action would be coded as a series of nerve impulses in the motor cortex, which would then be transmitted to the arm and hand muscles. These would variously contract and extend to produce the necessary movement. The motor processes convert information into action, which can be observed as the system's output (behaviour). There may be feedback processes built into the alarm technology to reinforce a correct response or to inform a reconsideration and a second attempt. Information processing models thus imply at least some serial processing and some discrete processing functions. They have been criticized by the connectionist school (see Best, 1992) because of this. The activity of the neural nets which comprise much of the cortex appears to involve parallel rather than serial processing and less differentiation of function than required by HIP models. Despite this, such models offer workable *explanations* of cognition and behaviour and there is still value to their development and exploitation.

It has been argued earlier in this book (see Chapter 4) that systems involving people are inherently more complex than purely engineering systems because of the complexity of human behaviour. The question here is that, given the view of the person as an information processing system, is behaviour complex because of the complexity of that system or because of the complexity of the environment (see Hollnagel, 1993)? Is the HIP system relatively simple or relatively complex? Simon (1969) is clear on this point:

> A man, viewed as a behaving system, is quite simple. The apparent complexity of his behaviour over time is largely a reflection of the complexity of the environment in which he finds himself (p. 25).

The HIP system, in Simon's (1969) view, is relatively simple and rule-driven and, within the framework of those rules, the cognitions at the heart of this system, are relatively reliable. Much effort has been expended, assuming the validity of this assertion, in discovering the 'rules' which drive human information processing (see, for example, the basic psychology texts produced by Gleitman, 1992 and Atkinson *et al.*, 1993, and the more advanced work of Best, 1992 on cognitive psychology). Human error has also been treated, within this framework, as rule-driven, and effort expended to uncover its 'rules' (see, for example, Reason, 1990, and Goodstein *et al.*, 1988).

Hollnagel (1993) has challenged the assertion that the HIP system is a simple one. He concludes, to the contrary of Simon's (1969) view, that man, viewed as a behaving system, is not simple and that the complexity and variety of the environment requires equally complex cognition in order for him to be able to cope. He argues that this is consistent with the notion of requisite variety as expressed in the so-named (cybernetic) Law of Requisite Variety (Ashby, 1956). The basic principle here is that, in order to control a system, the control function has to have at least as much variety as that system. Therefore, cognition, as the control function for coping with a complex environment, has itself to be at least as complex. Here the argument begins to touch on the basic observations of complexity theory (see, for example, Waldrop, 1992) which might offer a resolution. Can the argument be resolved by accepting that the interactions between even a small number of relatively simple rules can generate complex and, at times, seemingly unpredictable behaviour? If this is true then the HIP system, and cognition, might both be simple in their rule structures, yet complex in their output behaviour. Furthermore, the relationship between that complex behaviour and the complexity of the environment must be two-way. The success or failure of behaviour will not only shape the environment in some way but will also feed back to modify the rule structure that generated it and shape future behaviour. This chapter assumes the HIP model, and seeks to describe at an elementary level, what is known of its rule structure. However, it does so recognizing that that rule structure is capable of generating complex behaviour and that it is this complexity which makes systems involving people, themselves, inherently complex.

Various aspects of the HIP model are discussed in this section:

1 the sensory processes;
2 attention and perception;
3 cognitive processing, memory and attitudes;
4 skilled behaviour; and
5 feedback.

The sensory processes

Stimulation is received by the HIP system and converted to information in the form of a patterning of nerve impulses by the sense organs: eyes, ears, nose, mouth, skin, and muscle and joint receptors. The sense organs act as transducers, receiving information in a particular modality and converting it to the common currency of the nervous system – the patterning of nerve impulses. Sense organs change the patterning of nerve impulses in an already active system. They have specific sensitivities and threshold limits and may be subject to both natural or injury-related defects.

It has been estimated that 80 per cent of human knowledge is acquired visually (Hunter, 1992). Next to seeing, hearing is the most important sense used in information gathering. Physiological studies have determined that the ear can detect sounds which vary in frequency at about 20 cycles per second, or 1 hertz (abbreviated as Hz) to about 20 000 Hz (Hunter, 1992). However, it is important not to underestimate the value of the other senses in supporting the individual's ability to detect workplace hazards; for example, the sense of smell is important in detecting toxic substances.

The presence of hazards which are not perceptible to the senses, for example, gases such as methane, X-rays or ultrasonics, will not be detected by people unless suitable monitors and alarms are provided. Sensory defects may also prevent hazard information being received or distort it so as to make it uninterpretable. Similar effects may also be caused by interference with the reception process by the presence of equipment or protective clothing. Problems of this nature can be caused by the very equipment or clothing provided to protect people against exposure to danger. For example, hearing defenders provided to tree fellers may deprive them of the auditory cues provided by the changes in the noise of their saw as it cuts through the tree trunk which in turn enable them to estimate when the tree

Table 7.1 Guidelines for presenting 'work system' information (adapted from Hunter, 1992)

1   Use the most appropriate sensory channel

2   Use known physical limitations to ensure that information is presented clearly

3   Ensure that visual displays are arranged to minimize the amount of required eye movement for processing

4   Use visual displays (inter alia):
   a for pictorial or three-dimensional displays
   b for simultaneous comparisons
   c for fast choices among alternatives
   d for rapid scanning of material
   e at times when the ambient noise level is high

5   Use audio information transfer:
   a for fast two-way communication
   b for retention of familiar material
   c for caution, warning or alerting signals
   d at times when vision is limited

is about to fall. The loss of such information may impair decisions on taking evasive action. Guidelines for presenting 'work system' information so as to optimize processing of sensory information (Hunter, 1992) are included in Table 7.1.

Not all the external information which is available to the sensory processes or received by them is 'used' by the person, and that which is has to be 'interpreted' within the context of previous experience. These cognitive processes are usually described as 'perception' which, in turn, involves mechanisms of attention, pattern recognition and interpretation.

Perception

The person is confronted with a potentially bewildering barrage of stimulation from a vast array of different sources of information. Only some of this information is, however, taken in, interpreted and used. Information is selected in two ways:

1 peripherally by the nature and limitations of the person's sensory processes; and
2 centrally through the mechanisms of attention.

Generally speaking, there are two different types of model of attention (Best, 1992). The earlier type treated attention as the bottleneck in the HIP system. External stimuli could not be fully processed until they were attended to, and the attentional mechanisms were limited and could only handle a relatively small amount of information. More recent models treat attention as the mechanism which allocates processing resources. Here the function of attention is thus to bring other cognitive processes to bear on external stimuli so that information can be gathered about them and processed. An essential part of either system is the recognition of patterns present in the array of stimuli that bombard the senses.

At any time, people are conscious of various things taking place around them. In order to select what to attend to or to allocate processing resources appropriately, they must subconsciously process a wide array of information and then effectively reject much of it. This process could, in part, be peripheral but is more likely to be centrally driven. Much of it is automatic and unconscious. An interesting example of selective attention is provided by the so-called 'cocktail party' phenomenon. At a party, amid all the noise and clamour, you can often concentrate on what another person is saying and effectively 'cut out' the rest. However, if somebody else mentions your name in some far corner of the party, your attention may be suddenly drawn to them. In order for this to happen, you must have been monitoring and processing, albeit at a low level, much of the information that was not reaching consciousness, remaining ready to 'switch back in'.

The information that people take in is usually incomplete, ambiguous and, at the same time, context-dependent. People overcome this problem by actively 'interpreting' the available information in terms of their mental model of the world. There is an important role for past experience and memory in this process of interpretation. Past experience may be a very powerful influence. In some cases, the perceptions which are formed are shaped by the information that is 'expected' as much as that which is actually available. So, for example, a very brief exposure to a red ace of spades may result in the person describing that false playing card as an ace of hearts, or ignoring its unusual colour. The accident literature is replete with examples of this phenomenon. The Kegworth air disaster (8 January 1989) illustrates how the pilots' interpretation of the engine status and related control operation were shaped by their expectations and mental models and how these led them to shut down the wrong engine (Department of Transport, 1990).

Sensory information is interpreted in the context of the person's mental model of their world: it is adapted to that model and then accommodated within it. Therefore, perceptions are shaped by existing views of the world and, in turn, those views are modified in the light of the new information. Part of this process, and subsequent decision making, makes reference to information stored in memory and the higher level processing which follows is subject to the filtering and shaping implied by the existence of attitudes.

Decision making

The information that we attend to and then interpret contributes, with that stored in memory, to our mental model of the world. This model frames subsequent decision making (see Figure 7.1). People make decisions in several different ways (see Chapter 2) and the processes involved and the rules used are understood to some degree. For example, information can be processed in two ways, either subconsciously or consciously. Subconscious processing is automatic and gives rise to what is recognized as 'intuition'. Conscious processing, by contrast, appears to be more a logical step by step process. In both cases, some information is drawn from memory and much of the otherwise available information is incomplete for the purposes of logical decision making. This problem is worked around in various ways; for example, by using existing knowledge of:

1 similar situations;
2 frequently occurring situations; or
3 recent situations.

In making decisions about the design of safe systems of work in new and relatively unknown situations, managers might therefore base decisions on what is effective in other similar situations, what appears to be effective in most situations, or what was most effective in the last situation they had to deal with. These decision making strategies may or may not prove effective. Similarly, individual workers in an unfamiliar situation may base their behaviour on recent experience, although such experience might be wholly inappropriate. In doing so, they may have to rely on their ability to recall previous situations from memory.

Memory

Memory contributes to all aspects of cognition and involves at least three processes:

1 encoding;
2 storage; and
3 retrieval.

Information appears to be processed (or encoded) either verbally or iconically (by images) and storage may involve at least two sets of processes: (1) short term or working memory; and (2) long term or permanent memory.

Short term memory processes appear to have limited capacity and to lose information if it is not continually rehearsed or worked with. Short term memory appears to support conscious processing; and involve the serial processing of information. The person appears to act as a channel of limited capacity, but that limitation is defined in psychological rather than physical terms. Miller (1956) demonstrated that channel capacity is limited to between five and seven chunks of information. However, people can increase their channel capacity (and memory span) by deliberately 'chunking' information in ever larger amounts and giving those chunks particular meaning. A person can remember a string of about seven single meaningless digits, but also about seven ten-digit telephone numbers. Here the information is 'chunked' in a way that bestows both structure and meaning. A chunk is an organized cognitive structure: it can grow in size as more information is meaningfully integrated into it. Conversely, memory span may be dramatically reduced by the 'acoustic-similarity effect' (Conrad, 1964). The memory span for sound-alike items is markedly less than that for phonologically dissimilar items, even when the material is presented visually. Baddeley (1966) demonstrated that, whereas only 9.6 per cent of acoustically similar sentences could be recalled accurately from a list, 82.1 per cent of acoustically dissimilar sentences could be recalled.

The concept of 'limited capacity' is an important one for the HIP model, particularly in relation to problem solving. It means that the HIP system can be overloaded. It provides the requirement for both selective attention or the allocation of processing resources in order that a multitude of information processing tasks can be dealt with during one period of time. Naturally, mistakes and errors can be made if the wrong information is attended to or the wrong tasks are given processing priority. Human error is dealt with later.

Our memory of things can be affected by their position in a list or sequence of activities (Murdock, 1962). Memory is good for first things (primacy effect) and last things (recency effect) in a sequence, but is poor for middle of sequence information. This has implications for safety training. Training messages should be both clear and placed strategically in terms of their relative importance. The person's ability to hold information can also be affected if somewhat similar items are introduced before (proactive interference) or after (retroactive interference) the information to be remembered. The more similar the items are, the greater the interference. The sequencing of safety information in any presentation has, therefore, to be carefully planned.

Long term memory appears to involve schemata or structures for organizing and retaining information. These schemata may be represented in two different ways: as a set of hierarchical structures, in many ways similar to an active filing system (files within files), or as a neural net – a heterarchy of processing units. Processing within long term memory is probably parallel and unconscious. It gives rise to apparently instinctive decisions and requires low effort. Existing information structures may change information as it is stored (adaptation) and may themselves change as that new information is incorporated (accommodation). Information is consolidated into long term memory through use, meaning and importance.

It has been suggested that long term memory might be conceived, using the computer analogy, in terms of networks of memory elements; the elements are 'concepts' and are linked by their associations, or formal relationships. The concepts form the 'nodes' of such semantic networks and the relationships which link them are directional. For example, consider the fact that: *the man has safety glasses and is a safety adviser*. This is represented as a semantic network in Figure 7.2.

Searching and retrieving information from memory is held to be a search among the nodes (concepts) of the network. Anderson's (1976, 1983) ACT theory – adaptive control of thought – offers a detailed and systematized account of the application of network thinking beyond the question of memory to encompass most important aspects of cognition.

Information may be retrieved from long term memory by one of two processes: recall or recognition. These are different processes. For example, people can recognize others without necessarily being able to recall their names. Recognition is easier than recall because useful cues are often present in the environment, and such cues can be deliberately enhanced. Similarly, recall can be improved through the use of cognitive aids such as mnemonics (see below). Mnemonic devices are often incorporated into brand names, for example, Easy-Off oven cleaner.

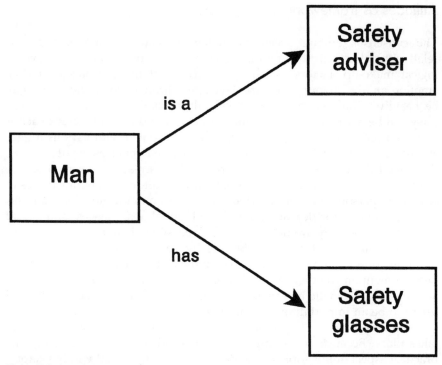

Figure 7.2 A semantic network

There are several different aids to memory, and information processing, which may be of importance in relation to the design of safe systems:

1 redundancy of information (this needs to be balanced with the need to prioritize essential information);
2 minimal interference;
3 meaningfully chunked information;
4 flagging of important information;
5 use of mnemonic devices;
6 minimal encoding requirements (dials say exactly what is meant by particular readings); and
7 provision of attentional devices.

Information retrieved from memory contributes to decision making which may, in turn, result in action often expressed as a change in verbal or locomotor behaviour. However, this decision making process is subject to the filtering and shaping implied by the existence of attitudes (see earlier).

Attitudes as frameworks for decisions

The consensus among social psychologists would define attitudes as relatively stable, but not immutable, components of the person's psychological make-up, factors which seemingly affect the person's cognition, emotion and behaviour. They are developed through experience, and may be heavily influenced by cultural, subcultural and local social pressures. They can be defined in terms of mini belief systems or tendencies to act or react in a certain (consistent) manner when confronted with various trigger stimuli. Allport (1935) has, for example, defined them as mental states, developed through experience, which are always ready to exert an active influence on an individual's response to any conditions or circumstances to which the person has been directed. The definition indicates that attitudes are 'developed' (or learnt) through individual experiences and social interactions (i.e. they are not innate). It is possible, therefore, that accident experience may affect attitudes. Studies by Leather and Butler (1983), for example, found that attitudes to safety in the construction industry differed depending upon whether the worker had personally experienced an injury/ accident. Workers who had personal experience of injury subsequently held safety to be an important aspect of their jobs.

Attitudes provide an important framework within which decision making takes place. Because of the importance attached to attitudes to safety, the practical aspect of this topic is also dealt with in some detail within Chapter 12. In that chapter the relationship between attitudes and behaviour is considered, and the issue of attitude change discussed. Explicit in many definitions of 'attitudes' is the notion that they are involved in determining the way the person thinks, feels and behaves in relation to particular situations or events. However, some authors have gone further and treated these three domains as fundamental components of an attitude: belief (the cognitive component), feeling (the affective component) and action (the behavioural component). Richardson (1977), for example, has defined an attitude as 'a predisposition to feel, think and act towards some object, person or event in a more or less favourable or unfavourable way'. The authors of this book treat attitudes as essentially cognitive with a possible emotional dimension. They recognize that while attitudes may influence behaviour (see Chapter 12), they cannot be defined in terms of behaviour. If they are, as some authors have stated, then it does not make sense to question the relationship between attitudes and behaviour because that relationship is established 'by definition'.

The cognitive and emotional dimensions of attitudes can be related to the 'hearts and minds' approach to safety discussed in Chapter 12. Each dimension can assume a positive or negative direction.

Valence is the term applied to the way in which the object of the attitude is evaluated – the degree of positive or negative direction. Valence is one of several characteristics of attitudes (commonly referred to in the literature – see, for example, Glendon and McKenna, 1995). Other characteristics include:

1 breadth – the number of attributes which characterize the object of the attitude, from very broad (for example, workplace safety) to very narrow (for example, a specific make of safety spectacles);
2 stability – how resistant the particular attitude is to change;
3 intensity – the strength of feeling, for example the holder of the attitude may have witnessed an horrendous accident and, thus, feel extremely strongly about 'safety';
4 salience – the degree to which an attitude occupies a person's awareness;
5 centrality – how much the attitude is part of an individual's self-concept (see Katz, 1960); and
6 behavioural expression – the degree to which our attitude is acted upon.

The architecture of attitudes is relatively well established by research (see, for example, S Cox, 1988). The question remains as to their function.

Frameworks and filters

By their very nature the population of attitudes that the person holds provides a framework within which decisions about people, objects, situations and events can be more readily made. For safety this would include plant and equipment, work processes, accidents or incidents and management commitment. Because attitudes are fairly persistent, they offer a relatively stable view of the person's world. The cognitive–emotional framework, which they combine to form, may facilitate decision making by providing a form of structure and predictability (see Chapter 12). In more rigid cases, this structure may even function as if it was a 'mental' analogue to a 'fixed action pattern'. However, this framework does not necessarily improve the accuracy of decision making. In some cases, it may lead an individual to adopt an unsafe practice – 'we have always done it this way, it is really very safe'. Beyond this facilitation by providing 'well worn' paths (strategies) or even actual decisions, the attitudinal framework may serve the person in other ways.

Katz (1960) has suggested that there are four major functions associated with attitudes, although it would appear that their validity has not been tested out to any great extent. Essentially what he has described are the goals or clusters of goals (and associated strategies) that attitudes serve. These are instrumental goals (maximization of reward), ego defence (self-protection), expression of identity and values (i.e. definition of self-concept) and knowledge acquisition. It would seem then that, according to Katz

(1960), attitudes facilitate decision making and in particular in relation to these four areas, all of which are, to an extent, person- (self-) centred.

In addition to providing a framework for decision making, attitudes may also serve as 'filters' and contribute to the higher processes of attention. Information and messages will be more readily accepted and adopted if they are perceived as consistent with existing attitude and belief systems, or act to reduce any inconsistencies in those systems. This is consistent with the notion of attitude structures as filters (see above). Inconsistency in a person's belief systems appears difficult to tolerate and can arise when the person behaves in a manner which does not correspond to their beliefs and attitudes. This may be particularly difficult for some types of person and acceptable to others (see Snyder and Kendzierski (1982), Chapter 12).

Control of motor processes and outputs

The output of the HIP system (see Figure 7.1) is often some form of behaviour, which might be verbal (what they say) as well as locomotor (what they do). Such behaviour is the result of the various cognitive processes described in the previous sections. Interestingly, while peoples' actions – locomotor behaviour – are the prime concern of most safety professionals, their verbal behaviour can also have an indirect but strong effect on safety. What people say contributes to the communication of information about, and attitudes to, safety, shapes expectations and can reward particular behaviour in others.

Central to the activities of social and occupational psychologists have been attempts to define the relationship between attitudes and behaviour (see, for example, Ajzen, 1991) and set out strategies for changing behaviour through attitude change. Although theoretical statements have generally assumed predictive and causal links, several empirical studies have found attitudes to be poor predictors of behaviour (see Chapter 12). Furthermore, it has been shown that changes in attitudes are not necessarily translated into corresponding changes in behaviour. However, it would appear that there is some correspondence between attitudes and behaviour but 'for some individuals more than others and for some situations more than others'. This has been demonstrated in the studies described in Chapter 12.

Learning and task performance

Much of everyday behaviour depends on the exercise of skills acquired through learning and perfected through practice. Such behaviour often takes the form of a sequence of skilled acts interwoven into well established routines, the initiation of each subsequent act being dependent on the

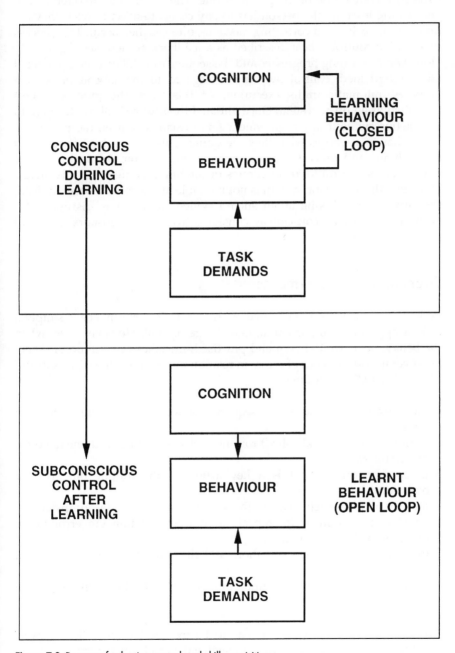

Figure 7.3 Process of adaptive control and skill acquisition

successful completion of the previous one. These routines are often learnt, and during learning the person has to pay close attention to what they are doing, monitoring and correcting mistakes, often as they occur. This process of adaptive control is best described as a feedback system (see Figure 7.3). However, as learning progresses and the sequence of skilled acts is perfected and strengthened, control over it is delegated to the subconscious as it becomes automatic in its execution. At this stage the process is best described as an open system. Once initiated the routine will run through to completion without any requirement for conscious control; the person no longer has to attend to what they are doing. This allows them to focus on other things and process other sources of information. Automaticity (the delegation of control to lower centres of the brain) confers real advantage. However, the automatic system is not infallible and errors do occur. Indeed, the very act of thinking about such a skilled routine may disrupt it; for example, if a person consciously considers 'walking downstairs' they will probably fall over.

Errors and task performance

Perhaps the most common source of error in task performance is simply a lack of appropriate information, knowledge or skill. However, even when the behaviour is well known and practised, mistakes can still occur. There are several *common* types of errors in relation to skilled routines (Reason and Mycielska, 1982), including:

1 the person selects the wrong sequence of actions or routine (a decision error);
2 a stronger (better established) routine replaces the intended one (an error of execution);
3 the person omits what is a key action in the routine (an error of omission);
4 the person loses their place in the sequence and either (a) jumps ahead, omitting an action or (b) repeats a completed action (an error of (a) omission or (b) commission);
5 the person forgets their intention, and the sequence stops (an error of inaction); and
6 the person has the right intention but 'works' with the wrong object (an execution error).

On occasion, routine behaviour has to be overridden and other 'newer' behaviours have to be acted out. This switch demands the person's attention and conscious control. Errors can also occur here if the person is distracted or attends to the wrong aspect of the situation. In the first case, more established but inappropriate routines may replace the new behaviour and, in the second, control over behaviour may be ineffective and the new behaviour may fail.

On the face of it, therefore, there are many ways (and many stages) at which even relatively simple tasks can be performed incorrectly. However, the reality is very different. People are reasonably good at detecting and correcting errors (for example). Reason (1990) has observed that error is neither as abundant nor as varied as its vast potential might suggest:

> Not only are errors much rarer than correct actions, they also tend to take a limited variety. Moreover, errors appear in very similar guises across a wide range of mental activities. Thus it is possible to identify comparable error forms in action, speech, perception, recall, recognition, judgement, problem solving, decision making, concept formation and the like (Reason, 1990, p. 2).

A concept which is useful in understanding human error and its classification is that of intention. Reason (1990) has described an algorithm for distinguishing the varieties of intentional behaviour (see Figure 7.4). The three main categories (Reason, 1990) are non-intentional, unintentional and intentional but mistaken actions. This algorithm provides a useful method of distinguishing between different kinds of intentional behaviour on the basis of yes-no answers to three questions:

1 were the actions directed by some prior intention;
2 did the actions proceed as planned; and
3 did they achieve the planned and expected result?

These questions may be used in an initial incident investigation or in a human error analysis to develop possible reliability models in which the relationships between intentional behaviours, necessary actions and preferred consequences are described. Such models may also be used to predict non-preferred consequences in hazard analyses (see Chapter 2) and can support the design of safe systems and the exercise of non-adaptive control.

Human error also needs to be considered in the context of particular tasks and task environments and in relation to the nature of the cognitive control involved. This is considered in the next section.

Skills, rules and knowledge

Human reliability specialists are concerned with examining human performance on specific tasks in order to predict and thus reduce the likelihood of human error. One of the leading authors in this area is Rasmussen and his skills–rules–knowledge (SRK) framework of 'cognitive' control has become

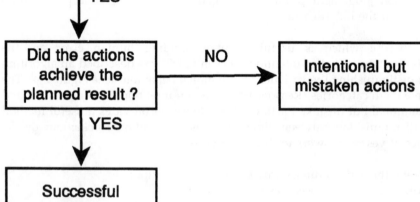

Figure 7.4 Algorithm for distinguishing intentional behaviours (adapted from Reason, 1990)

a market standard within the systems reliability community (Rasmussen, 1986). The SRK classification resulted from a longstanding concern with the problem of reducing errors in the control of complex systems. In Rasmussen's (1986) words, the

> model is not intended to be a psychological model of the mental processes of an operator in a basic task, but a functional model to illustrate general aspects of the operator's situation at a higher level as seen by the system designer; yet it has some similarities to a psychological model.

The SRK approach emerged as a result of an extensive review of incident and accident reports from nuclear power plants, chemical plants and aviation, and the observation that 'operator errors' only made sense when they were classified in terms of the mental operations underpinning them. The SRK model describes three distinct levels of performance (see Figure 7.5). Each level relates to decreasing familiarity with the environment or task. At the skill-based level, human performance is governed by stored patterns of preprogrammed instructions. It is characterized by 'free' and

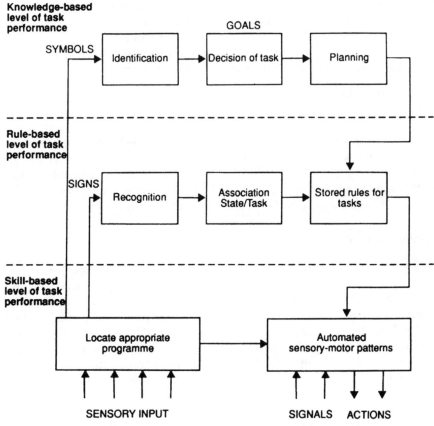

Figure 7.5 Rasmussen's SRK framework of task performance (from S Cox and Tait, 1991)

subconscious coordination between perception and motor actions (see Figure 7.3). The next level, the rule-based level, is applicable to tackling familiar problems in which solutions are governed by stored rules or procedures. The knowledge-based level comes into play in novel situations for which actions must be planned online, using conscious analytical processes and stored knowledge. With increasing expertise, the primary focus of control moves from the knowledge-based towards the skill based levels; but all three levels can coexist at any one time.

Reason (1985) has related his theory of errors to Rasmussen's 1986 model. For example, he argues that skill-based behaviour is distinguished from rule-based behaviour by being far more resistant to outside interference and far more subject to built-in control, which releases attention to other tasks. The 'design principles' behind these three levels of cognitive control are:

1 that control should be delegated to the most automatic level compatible with the predictability of the situation and with the repertoire of action sequences available to the individual facing it; and
2 that, as soon as problems are noted, control should be shifted up a level of consciousness in an attempt to solve the problem.

These principles represent a cognitive control version of subsidiarity, enhancing control at the most appropriate level for any particular situation or event.

This section of Chapter 7 has considered work behaviours in terms of learning and execution of particular tasks. However, work behaviours are also shaped by other factors including individual motivation. Motivating both individuals and groups of workers to behave safely has been the subject of much debate and is considered below.

Motivation

The terms 'motive' and 'motivation' have been applied to behaviour in three distinct ways (Buchanan and Huczynski, 1985): (1) to express the goals that people have; (2) to describe the cognitive processes that lead them to pursue particular goals; and (3) to consider the social processes through which individuals try to change the behaviour of others. Motives are primarily concerned with the goals of behaviour whereas motivation is the decision making process through which the individual chooses derived outcomes and sets in motion the necessary actions to achieve them (Buchanan and Huczynski, 1985).

There are various theories of motivation in management practice; these focus on either content or process. The former deal with the nature of the motivation, often based on an understanding of the person's needs (Maslow, 1954), while the latter describe the mechanisms by which motivational factors affect performance (for example, expectancy theory, see later). Maslow (1954) proposed five distinct categories of needs, arranged in a hierarchical order:

1 physiological needs (drives) – for example, hunger, thirst;
2 safety and security needs – in the sense of stability, dependency, protection, freedom from fear, anxiety, etc.;

3 belongingness needs – receiving and giving love and affection;
4 esteem needs – for strength, achievement, adequacy, confidence, independence *and* for reputation, recognition, attention, importance, etc.; and
  finally
5 self-actualization needs – for development of capability to the fullest
  potential.

The hierarchical structure leads to the 'prepotency principle'. According to
this principle individuals move upwards through this hierarchy by
satisfying a lower order need, which in turn leads to increased salience and
motivating force for attaining the next level. This principle has been
challenged (Alderfer, 1972). For example, some people may crave self-
actualization in preference to having the ostensibly lower order needs of
love satisfied.

There are two main problems associated with Maslow's theory; first, it
cannot easily be used to predict individual behaviour and second, it does
not provide a means of measuring motivation. Despite these criticisms
Maslow's (1954) approach can be applied as a framework for addressing
safety needs (see Table 7.2).

The mechanism through which outcomes become desirable and are
pursued is explained by expectancy theories of motivation. Vroom (1964)
produced the first systematic formulation of expectancy theory. His approach
also provided a way of measuring human motivation in terms of:

1 subjective probability (E), the individual's expectation that behaviour
  would lead to a particular outcome;
2 preferred outcome (P) (this may be positive, negative, or neutral); and
3 strength of the motivation to behave (F).

The strength of motivation or force to behave in a certain manner is
expressed by Vroom's (1964) expectancy equation:

$$F = E \times P.$$

In most circumstances a number of outcomes will result from a particular
behaviour. The expectancy equation thus has to be summed across all these
outcomes.

Expectancy and preferred outcome are multiplied because when either E
or P is zero, motivation is also zero. Thus if an individual believes that a
certain set of behaviours will lead to a particular outcome, but place no
value on the outcome, they will not be motivated to behave in that way.
Similarly, if a high value is placed on an outcome but the probability of
achieving it is zero, then motivation is zero. Only when both terms are
positive will motivation exist.

Table 7.2 Safety management: addressing motivation (Maslow's approach)

First Level	Physiological safety approach
	1 Threat of termination for violation of safety rules 2 Hard sell safety (i.e. blood and gore advertisements) 3 Scare procedures
Second Level	Safety and security safety approach
	1 Safety incentive plans (i.e. money) 2 Hard sell safety (i.e. blood and gore advertisements) 3 Negative safety audits (find something wrong) 4 Physical guarding rules and regulations 5 Scare procedures
Third Level	Belongingness and love safety approach
	1 Soft sell safety (i.e. identify safety with family and love) 2 Safety committees 3 Safety suggestion scheme 4 Recognition of individual safety awareness 5 Recognition of group safety awards 6 Positive safety
Fourth Level	Esteem safety approach
	1 Achievement orientated safety 2 Participative (by employees) safety 3 Safety suggestion schemes – recognition for contributors 4 Safety committees (employee-orientated) 5 Safety surveys (point out good and some poor points)
Fifth Level	Self-actualization safety approach 1 Creative safety 2 Participative (by employees) safety 3 Highest order achievement

The control and modification of safe behaviour often exploit such theories, either directly through the design of jobs (see Chapter 6) or through organizational procedures such as incentive schemes (see Chapter 12).

The acknowledgement that motivation and error are both part of the human condition has led to an enhanced consideration of these factors in the design of work systems. A question which has received less attention is the extent to which violations represent a risk to the integrity of such systems – what drives a worker to wilfully ignore safe working procedures? Reason (1990) defines violations as 'deliberate ... deviations from those practices deemed necessary ... to maintain the safe operation of a potentially hazardous situation'. The emphasis on the deliberate nature of such actions brings into play the individual's motivation (Vroom, 1964) in carrying out such behaviours. In terms of expectancy theory the worker obviously discounts the probability of potential harm and also places no value on safe methods in the successful completion of the task.

Clarke (1994) has argued for a systematic approach to violations and one which takes advantage of the move towards goal-orientated standards which logically follow on from organizational risk assessments. She maintains that prescriptive workplace rules (see Chapter 10) may create impossible situations for operators, where (in the extreme) deviations are necessary for system functioning, but may have severe consequences for the operator. Rules may thus be perceived as a 'trap' for employees, in which individual liability may be incurred if rules are violated, despite a culture which advocates company loyalty and getting the job done. Rule violations may be classified in terms of individual perceptions of the prevailing systems and culture (see Chapter 10) as part of the risk assessment process. They may also be considered in relation to their perceptions of risk (or danger). This is considered below.

Perception of risk

When people make judgements about things in the environment, whether it be a simple structural characteristic such as the size of an object or a more complex attribute like risk, they are swayed in those judgements by contextual information. This information can lead to illogical conclusions. Figure 7.6 (adapted from Boff and Lincoln, 1988) provides an example of the way in which people judge the size of objects using depth cues. It is known that as an object gets further away (judged by cues of depth) it appears to become smaller, although its actual size obviously has not changed. This effect can be manipulated to present the illusion shown in Figure 7.6. In an analogous way, people can be swayed in their judgement of risk by the various cues they receive with respect to a particular hazard or work environment. For example, if a group of workers have always shared a common access to their workstation with vehicles (such as forklift trucks) they may not perceive the potential danger associated with this practice. After all, it may be argued that if it was really unsafe they would not be allowed to use it (in practice, internal transport accidents are still a major problem (HSE, 1989, 1993)). Similarly, people are swayed in their perception of risk by the magnitude of the potential consequences of a particular activity (Slovic *et al.*, 1981). Thus, people feel more at risk when travelling by plane (because plane crashes kill hundreds) than when travelling by car (car crashes may only kill a handful), even if the actual (or objective) risk of being killed may be less.

Figure 7.6 Illustration of size (adapted from Boff and Lincoln, 1988). The perceptual processes are switched on to this illustration because of the perceived size as a consequence of depth cues.

There are a number of other factors which shape perception of risk and some of these are related to the general properties of a particular hazard–harm–risk relationship (as described in Chapter 2). These have been summarized in the Royal Society report on 'Risk' (1992) and are presented in Table 7.3 (adapted from S Cox and Tait, 1993). Pérusse (1980) has classified these factors in terms of two crucial dimensions, namely scope for human intervention (i.e. what could I do about this risk?) and 'dangerousness' (i.e. what is the worst thing that could happen as a result of this risk?). The latter may change as a worker becomes accustomed to a particular task. For example, the 'dangerousness' associated with the use of a particular piece of work equipment will not diminish over time; however, if an operator has not had an accident they may change their perception of the worst case scenario.

Table 7.3 General attributes of (hazard and) risk that influence risk perception

(1)	Voluntary exposure to hazard vs. involuntary exposure
(2)	Immediacy of effects of exposure vs. longer term effects
(3)	Uncertainty about probability and consequences of exposure
(4)	Catastrophic consequences are emotionally difficult to endure alongside consequence which threaten future generations (e.g. environment)
(5)	The degree of personal control over outcomes of exposure
(6)	Ignorance or lack of personal experience or understanding of hazard
(7)	Nature of benefits that may accrue (and to whom)
(8)	Risks associated with natural rather than man-made hazards
(9)	The nature and source of compensation for harms
(10)	Effects on children or vulnerable populations

Combining knowledge of perceptual processes in general, and of those specifically involved in workplace risk perception, allows some understanding of why individual perception of risk may be different from the objective risk. It has been further argued that objective risk, however calculated, cannot adequately account for error and behaviour in hazardous situations (T Cox and S Cox, 1993). That behaviour appears more strongly related to subjective or perceived risk because the person's perception of his or her world is his or her reality (Covello, 1983).

Perceptions of work related risks

The literature on perceptions of work related risks has been reviewed by the Human Factors Study Group of the Advisory Committee on Safety in Nuclear Installations (ACSNI, 1993). Table 7.4 includes examples of these studies and describes some of the key findings. In general, all of the studies are illustrative of the discrepancy between objective and subjective risk assessments and demonstrate predictable biases. For example, tractor drivers underestimated unusual risks (Singleton *et al.*, 1981) and overestimated more usual risks. Construction workers underestimated familiar risks (Zimolong, 1979), and nurses' perceived risk and related anxieties associated with HIV were distorted compared to those associated with HBV (Ferguson *et al.*, 1994).

Table 7.4 Examples of studies on workplace risk perception (source: ACSNI, 1993)

*Researcher(s)	Nature of study	Key findings
A Hale (1971)	*Light engineering workers.* Comparison between independent objective ratings of workplace risks with the operators' self-reported hazard awareness.	1 Identified critical role of worker experience of previous accidents, and of the frequency of accidents, in enhancing experience. 2 Evidence that some workers, regardless of accident experience, had greater appreciation of risks.
J Dunn (1972)	*Chainsaw operators.* Comparison of subjective risk with objective note derived from accident statistics.	1 Very little correlation between the two ratings.
B Zimolong (1979)	*Construction workers.* Perceived risk of falling accidents. Eight pictures of work situations presented, subjects asked to assess risks. Subjects chosen from six high risk occupational groups including carpenters, tile layers, scaffolding assemblers, construction workers, painters and steel construction workers.	1 Significant agreement within each group. 2 No relationship between objective risks and subjective assessment. 3 Risk which workers were most familiar with were underestimated.
O Ostberg (1980)	*Swedish forestry workers.* Nine hazards of tree felling operations illustrated and described in a booklet. Forestry workers were asked to make assessments of comparative severity.	1 High level of agreement on perceived relative riskiness of nine operations. 2 Substantial differences amongst six groups – fellers, trainers, safety officers, forestry school trainees, safety engineers and supervisors. 3 Trainers overestimated and supervisors underestimated risks.

Table 7.4 (Continued)

*Researcher(s)	Nature of study	Key findings
T Singleton et al. (1981)	*Tractor drivers.* Drivers were required to estimate the likelihood that a tractor would overturn in 10 different situations. Estimates were then compared with accident statistics (the accident statistics were judged to be reasonably reliable).	1 For two out of 10 ratings, there were substantial discrepancies. 2 Drivers seriously underestimated the risk of being hit by another vehicle on a public road (subjective ranking 10th; objective ranking 5th). 3 Drivers overestimated the possibility of a tractor rearing backwards whilst travelling up a steep slope (subjective risks 6th; objective risks 10th).
A Rushworth et al. (1986)	*Coal industry – Bunker maintenance workers.* 18 risky behaviours or situations were selected and descriptive scenarios devised. Experts ranked risks on 'anchored' scale and judgements of workers were then elicited.	1 Ranking exercise was valuable as precursor to informal discussions. 2 Respondents encouraged to reconsider rankings in the light of these discussions. 3 Useful tool to assess any uncertainties concerning workplace risks (trainees had high risk perceptions, together with uncertainty). 4 Experienced craftsmen and colliers (who carried out occasional bunker tasks) were less aware of hazards, neither did they show uncertainty. 5 Colliery bunker specialists had low hazard awareness and some uncertainty. However, area teams were high on confidence but underestimated hazards.
J Peterson et al. (1987)	*Forestry workers.* Natural experiment in shift from piecework payment to day rates.	1 Workers were well aware of safe procedures but found them to be too time consuming and a hindrance to earning. 2 Risk taking behaviours changed markedly when they went over to day rates.

Table 7.4 (*Continued*)

*Researcher(s)	Nature of study	Key findings
A Cheyne and S Cox (1994)	*Factory workers in a medium sized company. Managers, supervisors and machine operatives were asked to assess factory hazards and rank them. Ranking compared with objective accident and ill-health experience.*	1 Rankings of separate groupings were not identical and did not correspond with objective data with the exception of workplace noise. 2 Some workers in each of occupational groupings had greater appreciation of risk.
Ferguson *et al.* (1994)	*Nurses perception of risk associated with microbiological hazards. Psychometric analysis of vignettes and questionnaire studies.*	1 Perceived risk and related anxiety associated with HIV distorted compared to HBV and context- and knowledge-dependent.

Reactions to perceived risks

People differ in their reactions to perceived risks. Even when given the same information on workplace hazards, their responses vary. Such behavioural effects may result from differences in perception, experience, attitude, personality or skill.

Glendon and McKenna (1995) have identified the following factors which they believe affect individual reactions to perceived risks including:

1 individual differences – age, gender, personality;
2 perceived control by the person over the risk;
3 the person's existing motivations and displayed behaviours; and
4 the individual's 'set' or state of alertness – the extent to which they are able to receive information about the risk.

Ferguson and his colleagues (1994) have also demonstrated the importance of context and hazard knowledge in their work on nurses' perception of risk in relation to microbiological hazards.

The challenge for systems safety is to ensure effective management of both individual and collective reactions to perceived risks. This may be facilitated through various techniques including effective communication, organizational learning and training (see Chapter 12).

Summary

This chapter has considered the person as an active processor of information within the authors' model of safety management. It has discussed the psychology of safety in terms of individual cognition, emotion and behaviour. Issues of particular relevance to safety were introduced and are developed later in the text, including:

1 attitudes to safety and their possible relationships to safe behaviour;
2 human error – particular active failures associated with task performance;
3 safety motivations and the processes underpinning violations; and
4 risk perception.

None of these issues should be viewed in isolation because they naturally overlap as concepts and interact as processes. The individual is a key component in the authors' framework for understanding and managing safety, both as the focus of ultimate concern and as the agent of change both at the individual and organizational levels. As an agent of change, the individual acts both on their own behalf, looking after themselves, and as a key player in the shaping of their organization. Their actions in both respects depend on their perceptions, cognitions and motivations; these are set in the context of and affected by the individual's wider environment and, in particular, their jobs and organizations. The person effectively acts as an active information processor in this context, and can be adequately understood as such.

References

Advisory Committee on Safety in Nuclear Installations (ACSNI) (1993). *Third Report, Organising for Safety, of the Human Factors Study Group of ACSNI*, Suffolk: HSE Books.
Ajzen, I. (1991). 'The theory of planned behaviour'. *Organizational Behaviour and Human Decision Processes*, **50**, 179–211.
Alderfer, C.R. (1972). *Existence, Relatedness and Growth: Human Needs in Organizational Settings*. New York: Free Press.
Allport, G.W. (1935). 'Attitudes'. In *Handbook of Social Psychology* (C. Murchison, ed), Worcester, MA: Clark University Press.
Anderson, J.R. (1976). *Language, Memory and Thought*. Hillsdale, NJ: Erlbaum.
Anderson, J.R. (1983). *The Architecture of Cognition*. Cambridge, MA: Harvard University Press.
Ashby, W.R. (1956). *An Introduction to Cybernetics*. London: Methuen and Co.
Atkinson, R.L., Atkinson, R.C., Smith, E.E. and Bem, D.L. (1993). *Introduction to Psychology*. Forth Worth: Harcourt Brace Jovanovich College Publishers.
Baddeley, A.D. (1966). *Working Memory*. New York: Oxford University Press.
Best, J.B. (1992). *Cognitive Psychology*. New York: West Publishing Co.
Boff, K.R. and Lincoln, J.E. (1988). *Engineering Data Compendium Human Perception and Performance*. Harry G. Armstrong Aerospace Medical Research Laboratory, Human Engineering Division, Wright-Patterson Air Force Base, OH 45433, USA.
Buchanan, D.A. and Huczynski, A.A. (1985). *Organizational Behaviour, An Introductory Text*. Englewood Cliffs, NJ: Prentice Hall International.

Cheyne, A. and Cox, S. (1994). 'A comparison of employee attitudes to safety'. In *Proceedings of IVth Annual Conference on Safety and Well-Being at Work* (A. Cheyne, S. Cox and K. Irving, eds) Loughborough: Loughborough University of Technology.

Clarke, S. (1994). 'Violations at work: implications for risk management'. In *Proceedings of the IVth Conference on Safety and Well-Being at Work, November 1–2* (A. Cheyne, S. Cox and K. Irving, eds) 116–124, Loughborough, Loughborough University of Technology.

Conrad, R. (1964). 'Acoustic confusions in immediate memory'. *British Journal of Psychology*, **55**, 75–84, 291.

Covello, V.T. (1983). 'The perception of technological risk: a literature review'. *Technological Forecasting and Social Change*, **23**, 285.

Cox, S. (1988). *Employee Attitudes to Safety*. M.Phil. thesis. Nottingham: University of Nottingham.

Cox, S.J. and Tait, N.R.S. (1991). *Reliability, Safety and Risk Management*. London: Butterworth-Heinemann.

Cox, S.J. and Tait, N.R.S. (1993). 'From risk analysis to risk management – the developing role of the engineer'. In *Engineers and Role Issues, Proceedings of the Symposium of the Safety and Reliability Society*, October, Altrincham, UK.

Cox, T. and Cox, S. (1993). *Psychosocial and Organizational Hazards: Monitoring and Control*. European Series in Occupational Health no. 5. Copenhagen, Denmark: World Health Organization (Regional Office for Europe).

Department of Transport (1990). *Report on the Kegworth Air Disaster*. London: HMSO.

Dodd, D.H. and White, R.M. (1980). *Cognition: Mental Structures and Processes*. Boston, MA: Allyn and Bacon.

Dunn, J.G. (1972). 'Subjective and objective risk distribution'. *Occupational Psychology*, **46**, 183–187.

Ferguson, E., Cox, T., Farnsworth, W. and Irving, K. (1994). 'Nurses' anxieties about biohazards as a function of context and knowledge'. *Journal of Applied Social Psychology*, **24**(10), 926–940.

Gleitman, H. (1992). *Basic Psychology*. New York: Norton and Co.

Glendon, A.I. and McKenna, E.F. (1995). *Human Safety and Risk Management*. London: Chapman and Hall.

Goodstein, L.P., Anderson, H.B. and Olsen, S.E. (1988). *Tasks, Errors and Mental Models*. London: Taylor and Francis.

Hale, A. (1971). *Appreciation of Risk at Work. 2000 Accidents. National Institute of Industrial Psychology (Report No. 21)* (P.I. Powell, M. Hale, J. Martin and M. Simon, eds), London: NIIP.

Health and Safety Executive (1989). *Human Factors in Industrial Safety*. London: HMSO.

Health and Safety Executive (1993). *Transport Kills*. London: HMSO.

Hollnagel, E. (1993). *Human Reliability Analysis: Context and Control*. London: Academic Press.

Hunter, T.A. (1992). *Engineering Design for Safety*. USA: McGraw-Hill.

Katz, D. (1960). 'The functional approach to the study of attitudes'. *Public Opinion Quarterly*, **24**, 163–204.

Leather, P.J. and Butler, A.J. (1983). *Attitudes To Safety Among Construction Workers – A Pilot Survey*. Report prepared for the Building Research Establishment by the Department of Behaviour in Organisations, University of Lancaster.

Maslow, A.H. (1954). *Motivation and Personality.* New York: Harper.

Miller, G.A. (1956). 'The magical number seven plus or minus two: some limits of our capacity for processing information'. *Psychological Review,* **63**, 81–97.

Murdock, B.B. Jnr (1962). 'The serial position of free recall'. *Journal of Experimental Psychology,* **64**, 482–488.

Ostberg, O. (1980). 'Risk perception and worker behaviour in forestry: implications for accident prevention policy'. *Accident Analysis and Prevention,* **12**(3), 189–200.

Pérusse, M. (1980). *Dimensions of Perception and Recognition of Danger.* Ph.D. Thesis, Aston University, Birmingham.

Peterson, J.M., MacDonell, M.M., Haroun, L.A. and McCracken, S.H. (1987). *Expediting Cleanup at the Weldon Spring Site Under C.E.R.C.L.A. and N.E.P.A.*

Rasmussen, J. (1986). *Information Processing and Human-Machine Interaction: An Approach to Cognitive Engineering.* Amsterdam: Elsevier.

Reason, J.T. (1985). 'Generic error modelling system (G.E.M.S.); a cognitive framework for locating human error forms'. In *New Technology and Human Error* (J. Rasmussen, J. Leplat and K. Duncan, eds), Chichester: Wiley & Sons.

Reason, J.T. (1990). *Human Error.* Cambridge: Cambridge University Press.

Reason, J.T. and Mycielska, K. (1982). *Absent Minded? The Psychology of Mental Lapses and Everyday Errors.* Englewood Cliffs, NJ: Prentice-Hall.

Richardson, A. (1977). 'Attitudes'. In *Introductory Psychology* (J.C. Coleman, ed), London: Routledge & Kegan Paul.

Royal Society (1992). *Risk: Analysis, Perception and Management.* Report of a Royal Society Study Group. London: The Royal Society.

Rushworth, A.M., Best, C.F., Coleman, G.J., *et al.* (1986). *Study of Ergonomic Principles in Accident Prevention and Bunkers.* Institute of Occupational Medicine Report No. TM/86/5, Final Report on CEC Contract No. 7247/12/049, IOM, Edinburgh.

Simon, H.A. (1969). *The Sciences of the Artificial.* Cambridge, MA: MIT Press.

Singleton, W.T., Hicks, C. and Hirsch, A. (1981). *Safety in Agriculture and Related Industries.* Department of Applied Psychology Report No. AP 106, University of Aston, Birmingham.

Slovic, P., Fischhoff, B. and Lichtenstein, S. (1981). 'Facts and fears: understanding perceived risk?' In *Societal Risk Assessment. How Safe is Safe Enough?* (R.C. Schwing and W.A. Albers, eds) 181–214, New York: Plenum Press.

Snyder, M. and Kendzierski, D. (1982). 'Acting on one's attitudes: procedures for linking attitude and behaviour'. *Journal of Experimental Social Psychology,* **18**, 165–183.

Vroom, V.H. (1964). *Work and Motivation.* Chichester: Wiley.

Waldrop, M.M. (1992). *Complexity.* New York: Simon and Schuster.

Zimolong, B. (1979). 'Risikoeinschatzung und unfallgefahrdung beim rangieren'. *Eitschrift für Verkehrssicherheit,* **3**, 109–114.

8
Work-related stress

Introduction

One of the outcomes of a failed work system may be the experience of stress by some or all of those involved with that system. In systems theory terms, the experience of stress through work may result from failures of both non-adaptive and adaptive control; failures of design and of management.

It has been widely and variously suggested that the experience of work-related stress, and its psychophysiological and behavioural correlates, may threaten availability for work, the safeness and effectiveness of work behaviour, the quality of working life and both psychological and physical health. This chapter considers the nature of work-related stress, its origins and its possible effects at the individual and organizational levels. It discusses the part played by the experience of stress in linking exposure to work hazards to their effects on the individual and the organization. In doing so, it draws on the transactional model of T Cox and his colleagues (T Cox, 1978, 1985; T Cox and Mackay, 1981; T Cox and Ferguson, 1991; T Cox and Griffiths, 1995) and interprets this model within the framework suggested in this text for the management of safety.

Hazards and the experience of stress

As suggested in Chapter 2, work hazards can be broadly divided into the more physical and tangible, which include the biological, biomechanical, chemical and radiological, and the psychosocial. While most people understand something about the nature of the first broad category of hazards, the concept of a psychosocial hazard is more difficult to grasp. The International Labour Office (ILO) (1986) has attempted a definition. It has suggested that psychosocial hazards are those which relate to the inter-actions among job content, work organization, management systems, environmental and organizational conditions, on one hand, and workers' competencies and needs on the other. Those interactions which prove hazardous influence health through workers' perceptions and experience, a point which is explained more fully below.

The available scientific evidence suggests that exposure to work hazards is associated with the experience of stress, job dissatisfaction and ill-health. This evidence has been variously reviewed over the last two decades by a large number of individual researchers and research and policy organiza-tions (for example, Levi, 1984; Cooper and Marshall, 1976; T Cox, 1993; Levi et al., 1986; Warr, 1992; National Institute of Occupational Safety and Health (NIOSH), 1988; T Cox and S Cox, 1993). As a summary of all the reviews, Table 8.1 (derived from T Cox, 1993) outlines nine different characteristics of jobs, work environments and organizations which are hazardous. They relate to aspects of organizational function and culture, participation/decision latitude, career development, role in organization, job content, workload/work pace, work schedule, interpersonal relationships at work and work–home interface. Under certain conditions (also noted in Table 8.1), each of these nine characteristics of work has proved stressful and/or harmful to health.

Table 8.1 might be used as the basis for a checklist for identifying psychosocial hazards. Such a checklist could offer a useful starting point for hazard identification, inspection and monitoring. The present authors have used Table 8.1 in this way in developing a system for the assessment of work-related stress in hospital based nursing (T Cox et al., in press). A somewhat similar approach has been taken in Finland where the Institute of Occupational Health has published guidance, in the form of such a checklist, for the *Assessment of Psychic Stress Factors at Work* (Elo, 1986). This checklist concerns working conditions, work organization, job content, and social factors much as described in Table 8.1.

Karasek (1979) has drawn attention to the possibility that such work characteristics may not be simply (linearly) associated with health, but that they might combine interactively in relation to their health effects. For example, analysing data from the USA and Sweden, he found that workers in jobs perceived to have both low decision latitude and high job demands were particularly likely to report poor health and low satisfaction. Later studies appeared to confirm the theory. For example, a representative sample of the male Swedish workforce was examined for depression, excessive fatigue, cardiovascular disease and mortality. Those workers whose jobs were characterized by heavy workloads combined with little latitude for decision-making were represented disproportionately on all these outcome variables. The lowest probabilities for illness and death were found among work groups with moderate workloads combined with high

Table 8.1 Psychosocial hazards of work (adapted from T Cox, 1993)

Work characteristic	Hazardous conditions
Organizational function and culture	Poor communications Organization as poor task environment Poor problem-solving environment Poor development environment
Participation	Low participation in decision making
Career development and job status	Career uncertainty or career stagnation Poor status work or work of low social value Poor pay, job insecurity or redundancy
Role in organization	Role ambiguity or role conflict Responsibility for others or continual contact with other people
Job content	Ill-defined work or high uncertainty Lack of variety Fragmented or meaningless work Under utilization of skill Physical constraint
Workload and work pace	Quantitative work overload or underload Qualitative work overload or underload High levels of pacing Lack of control over pacing Time pressure
Work organization	Inflexible work schedule Unpredictable hours Long or unsocial hours Shift working
Interpersonal relationships at work	Social or physical isolation Lack of social support Interpersonal conflict and violence Poor relationships with superiors
Home–work interface	Conflicting demands of work and home Low social or practical support from home Dual career problems

control over work conditions (Karasek, 1981; Ahlbom *et al.*, 1977; Karasek and Theorell, 1990; Karasek *et al.*, 1981).

The combined effect of these two work characteristics is often described as a true interaction but despite the strong popular appeal of this suggestion there is only weak evidence for its support (Kasl, 1989; Warr, 1990).

Karasek's (1979) own analyses suggest an additive rather than a synergistic effect, and he has admitted that 'there is only moderate evidence for an interaction effect, understood as a departure from a linear additive model'. Simple additive combinations have been reported by a number of researchers, for example, Hurrell and McLaney (1989), Payne and Fletcher (1983), Perrewe and Ganster (1989) and Spector (1987). Whether or not perceived job demands and decision latitude combine additively or through a true interaction, Karasek's work clearly shows that they are important factors determining the effects of work on workers' health.

Hazards, stress and harm

In Chapter 2, it was argued that there are, at least, two pathways which link exposure to the hazards of work to the harm that they might cause: first, a direct physicochemical mechanism, and second, a psychophysiological stress-mediated mechanism. These mechanisms were summarized in Figure 2.1. Several important points were noted about this simple model: the relevant points are summarized here. First, the two mechanisms do not offer alternative explanations of the hazard–health relationship; in most hazardous situations both may operate and interact. Additive and synergistic interactions are possible. Second, many of the effects of psychosocial hazards are undoubtedly mediated by psychophysiological – stress related – processes; indeed the definition of a psychosocial hazard offered by the ILO (1986) makes this inevitable. Third, distress related to exposure or threat of exposure to physical hazards can affect health through both psychophysiological and behavioural pathways. Fourth, the effects of stress on health are not confined to an impairment of psychological and social wellbeing but can also impact on physical wellbeing.

Essentially, the model presented in Figure 2.1 positions the experience of stress as a 'bridge' between hazard exposure and harm. Given present knowledge, it is possible to describe the nature of this bridge in more detail. The experience of stress is usually accompanied by changes in both physiological function and behaviour. Both may have implications for health. For example, stress-related changes in endocrine activity may contribute to the development of a wide range of physical pathologies (see below). At the same time, stress-related increases in health risk behaviours, such as smoking, or decreases in health promoting behaviours, such as exercise, may also contribute to pathology. The impairment to health associated with the experience of stress may, in turn, feed back and reduce the person's ability to cope both with the on-going situation or any new demands and challenges. This feed back may serve to increase the person's vulnerability to stress. The model present in Figure 2.1 is developed further in Figure 8.1.

It should be clear from the preceding discussion, and the model presented in Figure 8.1 that work-related stress is not simply or solely an occupational mental health issue. The experience of stress through work can arise from exposure to both the more tangible and physical hazards of work, as well as the psychosocial hazards, and it can harm physical health as well as psychological health. Work-related stress is an occupational health and safety issue in the full sense of the term.

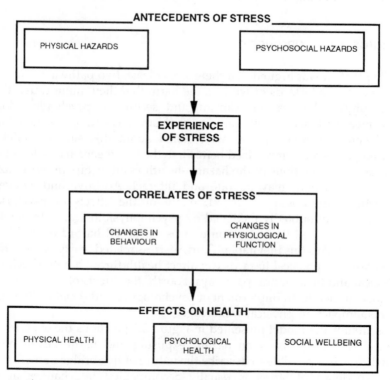

Figure 8.1 The antecedents, correlates and effects of stress

The antecedents of the experience of work-related stress lie partly in the characteristics of work, and the context to work. However, these hazardous characteristics – they carry the potential for harming the health of the individual – are only part of the overall 'stress' equation. They represent only one factor in that equation. The 'stress equation' is a simple and symbolic model of the decision architecture which underpins the cognitive appraisal process and the stress state. However, while a general equation can be described (see below), the particular combination of factor values which is associated with the experience of stress is very much individually driven. There are marked individual differences in relation to the experience of stress through work.

Cognitive appraisal: the architecture of the stress state

There has been much debate over the definition of stress, and some questioning of the usefulness of the concept. The continuation of this discussion, and the arguments which characterize it, is now largely fuelled by relative ignorance of the scientific literature and by dogmatism. To those who are informed two things are clear: first, that the concept of work-related stress offers a valuable economy in our description of the mechanisms which link the exposure of workplace hazards to the harm that they can cause, and, second, within this framework, there is a reasonably strong consensus of the nature of the processes involved with the stress concept and on the nature of the concept itself. True there are three different types of approach discernible in the scientific literature (see, for example, T Cox, 1978, 1993) but these simply offer different views or perspectives on the same overall process. They are similar, by analogy, to describing a car from the front, from the back and in terms of its function and behaviour. There is just one car but three different views of it. The consensus which emerges from the literature treats stress as a particular psychophysiological state, one which occurs as a result of exposure to aversive, noxious or otherwise threatening aspects of work – the hazards of work. This state appears to be characterized by extremes of arousal and by feelings of distress, and, sometimes, helplessness. It is strongly correlated with physiological and behaviour changes; many aspects of which are health-related. In one sense, stress is the negative emotional reaction to work which is difficult to cope with.

Work-related stress is a complex psychophysiological state deriving from the person's cognitive appraisal (see Figure 8.2) of the extent to which they can cope with the demands of the (work) environment (T Cox, 1978). Many believe that stress exists in the person's recognition of their inability to cope with the demands of the (work) situation and in their subsequent experience of distress. Stress is thus not an observable or discrete event. It is not a physical dimension of the environment, a particular piece of behaviour, or a pattern of physiological response. It has been suggested (T Cox, 1978) that the process of appraisal takes account of at least four factors:

1 the demands on the person, matched against
2 their ability to meet those demands (personal resources);
3 the constraints that they are under when coping; and
4 the support received from others in coping.

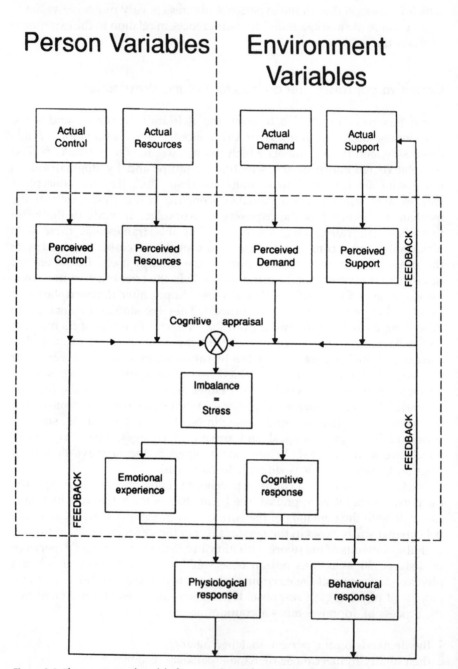

Figure 8.2 The transactional model of stress (adapted from T Cox, 1978)

This process of cognitive appraisal and its outcomes are represented diagrammatically in Figure 8.2. Together the four factors define the stress equation:

stress state = f (perceptions of (demand, ability to cope, control, support)).

There is sufficient evidence to allow speculation over the way in which the four factors combine. It is hypothesized here that the balance between perceived demand and perceived ability to cope is central to the equation, and that this balance interacts with the level of perceived control while perceived support contributes in an additive manner. Several other authors have suggested similar models. A classic stressful situation would involve work demands which are not well matched to the knowledge and skills of workers or their needs, especially where those workers have little control over work and receive little support at work (Payne, 1979; T Cox and Mackay, 1981; T Cox, 1990; T Cox and Ferguson, 1991).

Expanding this argument, the absolute level of demand would not appear to be the important factor in determining the experience of stress. The discrepancy that exists between the level of demand and the person's ability to cope (personal resources) would appear to be more important. Within reasonable limits a stress state can arise through overload (demands>abilities) or through underload (demands<abilities). It has been added that a state of stress may exist only if the person believes that the discrepancy is significant.

The existence of stress is often signalled to the person through the experience of negative emotion – unpleasantness and distress. Both the experience of stress and later attempts at coping may be detrimentally affected if the person's actions are constrained, or if little support is received from others.

Common confusions

A number of common confusions exist, the most pernicious misunderstandings are the relationships between stress and arousal, and stress and demand. They result in the belief that some stress is 'good for the individual'. This belief is often used to excuse poor management and a lack of attention to health and safety issues.

The term 'arousal' refers to the person's level of wakefulness, activation or energy utilization and alertness. Levels of arousal are independent of the experience of pleasantness or unpleasantness (distress) that accompany them. High arousal may be pleasant or unpleasant, as may low arousal. Somewhat similarly distress (unpleasantness) might be accompanied by either high or low arousal. Stress is the combination of extremes of arousal (high or low) with feelings of distress (extreme unpleasantness). This model of mood has been captured in the Stress Arousal Checklist (SACL: Mackay *et al.*, 1978; T Cox and Mackay, 1985) which offers two orthogonal scales measuring arousal (alertness, wakefulness and levels of activity) and stress (unpleasantness, tenseness and distress). This checklist has been used internationally in a wide variety of different research projects both in occupational health and safety, and, more generally, in psychology.

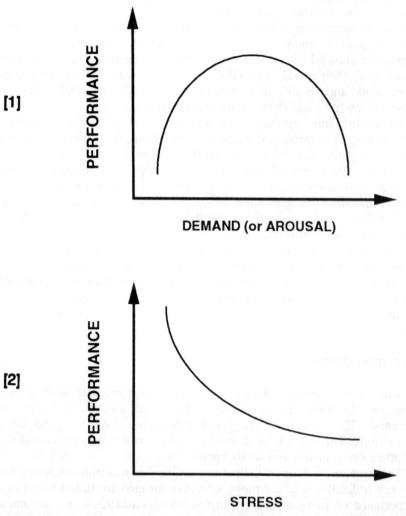

Figure 8.3 Relationships between (1) demand (or arousal) and performance and (2) stress and performance

Stress occurs when the person cannot easily cope with the demands of work. Demand in itself is not inherently stressful. Indeed, a certain level of demand is necessary to ensure adequate levels of arousal, wakefulness and activity, learning and satisfaction. It is when demand does not easily match the person's ability to cope, often when demands are too great (overload) but also when they are not great enough (underload), that the person experiences stress.

The empirical data suggest that there is an inverted U-shaped relationship between arousal and performance, and between demand and performance, but a negative and monotonic relationship between stress and performance (see Figure 8.3).

The effects of stress

The experience of stress can detrimentally affect the way a person feels, thinks and behaves. The effects of stress may be expressed in various ways: through feelings of distress, increased irritability, poor decision-making, excessive smoking and drinking, poor diet, impaired sleep and sexual behaviour (these two being particularly sensitive to stress and anxiety), inadequate exercise and an inability to relax. Stress may also produce changes in physiological function.

The emotional experience of stress

Kagan (1975) argued that there is no emotional experience or mood state diagnostic of stress. Rather stress is represented at different times and in different people by a wide range of different negative emotions and

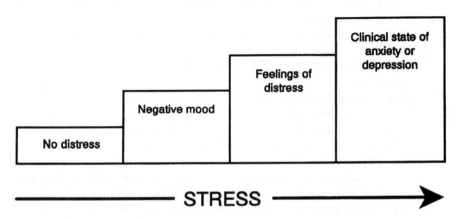

Figure 8.4 The experience of stress

combinations of emotion. Collectively, this kaleidoscope of feeling has been referred to as distress. Research by T Cox and Ferguson (1991), exploring the model of stress appraisals suggested by Lazarus and Folkman (1984), has suggested that most people identify stressful situations largely in terms of them being 'anxiety-producing' or 'depressing'. Earlier research by Mackay, T Cox and their colleagues (Mackay *et al.*, 1978; T Cox and Mackay, 1985) has attempted to capture the concept of 'distress' in terms of mood describing adjectives, and has defined 'stress' in these terms in relation to feelings of unpleasantness or negative hedonistic tone. Findings such as these allow a more precise statement on the nature of stress as an emotional experience (see Figure 8.4).

Several well developed psychometric instruments are available to assess different regions of this 'distress' dimension by self-report.

Behaviour under stress

Predicting people's behaviour under stress has been a prime concern for both psychologists and managers alike. While there is no reliable formula for exactly predicting what any one person will do, general principles can be described.

1 In a situation of severe stress, the person may attempt to escape from that situation, become aggressive or freeze. The structure of the situation may help determine which response occurs. Should the person remain still to avoid being noticed? If noticed can they physically escape? Are they cornered? Should they attempt to defend themselves?
2 In less stressful situations, the person will decide on their main task (among the many they are engaged in) and focus down and concentrate on it. Performance of this task may actually improve while that on other tasks may be impaired.
3 The person may defend against admitting the stressful problem; they might deny it – 'there is nothing wrong'. This will prevent them from dealing with the situation. This process of denial might be much stronger in a (work) group – group think.
4 Some people will react to stressful challenge by simply working harder and longer until they are eventually exhausted. They attempt to maintain performance against increasing demands by increasing their effort.
5 How the person reacts will depend on a number of factors, such as whether they work as part of a team, how committed they are to their social group and organization, and how they are managed. It may also depend on the nature of the task. Table 6.5 has illustrated a number of performance shaping factors which are used to support determinations of human error probabilities.

A final point of importance relates to the psychological aftermath of stress. Many people who experience severely stressful situations will be affected by that experience months, even years, later. Such post-traumatic reactions may explain unexpected deterioration in personal health and performance. The immediate management of post-stress situations may determine if the affected person suffers in the longer term.

Coping

Attempting to cope – coping – is a natural response to experiencing stress: it is a form of problem solving behaviour. It is important to understand that coping denotes a particular type of response and not everything that follows from the experience of stress. It may involve both cognitive and behavioural strategies and represents either an adjustment to the situation or an adjustment of the situation. Coping may be successful because the source of the problem has been dealt with (direct action) or because the experience of stress has been directly reduced (palliation). Whether successful or not, the events implied by coping will feed back to alter the person's initial appraisal of the (work) environment and, as a result, of other aspects of the overall process.

Physiological effects of stress

Since the early work of Cannon (1931) and Selye (1950, 1956) much of the physiology of stress has focused on two neuroendocrine systems: the sympathetic-adrenal medullary system and the hypothalamo-anterior pitui-tary-adrenal cortical system. In a sense both these systems focus on the function of the adrenal glands, and there are several reviews of the role of the adrenals in stress physiology (for example, Selye, 1950; T Cox and S Cox, 1985; Szabo et al., 1983). Recent studies in psychophysiology have extended beyond consideration of these two systems to psycho-neuro-immunology. For many, these changes will simply represent a modest dysfunction, discomfort and an impaired quality of life. For others they might be translated into poor performance at work, into other psychological and social problems, and even into poor physical health. Behavioural changes which are known to affect health, combined with changes in endocrine and immunological function, may describe the two pathways or channels by which the psychological experience of stress translates into more tangible physical pathology such as an increased likelihood of early coronary ill-health (T Cox et al., 1983).

Health

The changes which accompany the experience of stress may have very different immediate and long term consequences. For example, in introducing the idea of 'diseases of adaptation', Selye (1950) contrasted the short term adaptiveness of the neuroendocrine stress response with its longer term involvement in pathogenesis. The phytogenesis of these responses has depended on their immediate value to survival behaviour but, beyond that, they may extract a cost to health. It now appears that stress may contribute to ill-health in at least two ways. Firstly, the experience of stress may be associated with changes in attitudes and behaviours which relate to the maintenance of a healthy state. These changes may involve either the inhibition of health promoting behaviours, such as exercise and the practice of relaxation, or they may involve the development of health threatening behaviours, such as smoking or excessive drinking. Secondly, the neuro-endocrine responses to stress may interfere with normal physiological function and either inhibit the body's natural defences or promote pathogenic change.

The wider effects of stress involve changes in the function of systems in which the individual is but a component; these may be the person's social group, work organization or the services offered and demanded by the community or society at large. With regard to work, the changes of interest can be loosely ordered in terms of how directly they relate to individual behaviour. At one end of this continuum there are changes in sickness absence, timekeeping and task performance. At the other end, labour turnover and group morale, productivity and industrial relations are less directly dependent on the behaviour of any one person.

Harm to the organization

If key workers or significant numbers of workers experience and express the effects of stress at work, the problem assumes organizational proportions. There has been some suggestion that if about 40 per cent of workers in any group (department or organization) experience stress-related problems, that group or organization can also be said to be unhealthy in some way. The possible effects of stress of more direct concern to organizations appear to be the following:

1 reduced availability for work involving high turn-over, absenteeism and poor time-keeping – all essentially escape strategies;
2 impaired work performance and productivity: quantity and quality;
3 increased unsafe behaviour, near-miss and accident rates; and
4 increase in complaints from clients and customers.

For some, escapist strategies may not be personally or professionally acceptable: people may continue to turn up for work under stress but perform poorly.

The various effects of stress both on the individual and the organization are summarized in Table 8.2.

Table 8.2 Harm – possible effects of stress

Possible physical effects

Allergies
Arrhythmias
Backaches
Cancers
Chest pain
Colds and flu
Colitis
Constipation and diarrhoea
Cystitis
Dermatitis
Digestive problems
Dizziness
Headaches
Heart disease
Impaired sleep
Injuries
Menstrual problems
Migraine
Mouth ulcers
Obesity
Palpitations
Respiratory problems
Sexual problems
Skeletomuscular problems
Strokes
Sweating
Trembling

Possible psychological effects

Addictions and dependencies
Boredom
Depression
Eating disorders
Inability to concentrate
Insomnia
Irritability
Low self-esteem
Mental breakdown
Nervousness
Obsessions
Phobias
Post-traumatic stress disorder
Suicide
Tiredness
Tremor and stuttering

Possible social effects

Apathy
Family breakdown
Impulsive behaviour
Interpersonal aggression
Poor interpersonal relationships
Social isolation
Marital breakdown

Possible work-related effects

Absenteeism
Poor timekeeping
High turnover
High rate of client/customer complaints

Poor concentration
Poor quality of work
Increase in unsafe behaviour and accident rates

Vulnerability

Individual differences are obvious both in exposure to psychosocial hazards, and in workers' responses to them. In particular there has been much research into vulnerability to stress (cf: type A behaviour: Friedman and Rosenman, 1974) and resistance to stress (hardiness: Kobasa, 1979; Kobasa and Puccetti, 1983; Kobasa et al., 1981, 1982). However, these individual differences may be secondary to or, at least, combine with different sets of circumstance to create vulnerable groups of workers. Several different reviews have identified possible vulnerable groups (see, for example, Levi, 1984) including young workers, older workers, migrant workers, handicapped workers and women workers. Group differences may represent the effects of individual differences which are common to and characteristic of particular groups, or the effects of common patterns of exposure to hazardous work conditions (or some combination of the two). Determining exposure is not sufficient in itself to warrant labelling a particular group vulnerable or 'at risk', the health effects of such exposure have also to be demonstrated.

Control of psychosocial hazards and work-related stress

It has been argued that work-related stress is an occupational safety and health issue. It would therefore be logical to attempt to solve stress-related problems using those strategies proved successful in other areas of health and safety. At present, the dominant paradigm (see Chapters 2 and 9) is that of risk management. It has been variously suggested by the authors (T Cox et al., 1990; T Cox and S Cox, 1993; T Cox, 1993; T Cox and Griffiths, 1995) that not only does the risk assessment – risk management paradigm offer an effective way of dealing with the more tangible and physical hazards of work, but it may also be effective in relation to psychosocial hazards. A particular account of the cycle of control offered by this paradigm for work-related stress is elaborated in Figure 8.5. This is taken from T Cox and Griffiths (1995).

T Cox et al. (1990) have suggested that at least three levels of control strategy have been adopted by organizations to deal with psychosocial hazards, stress, and their health effects: preventive strategies, often control by design or through worker training, to remove the hazard or reduce its impact on workers or their likelihood of exposure; strategies involving timely reaction (see T Cox, 1993), often based on management and group problem solving, to improve the organization's ability to quickly recognize and deal with problems as they arise; and rehabilitative strategies, often involving enhanced employee support (including counselling) to help workers cope with and recover from problems.

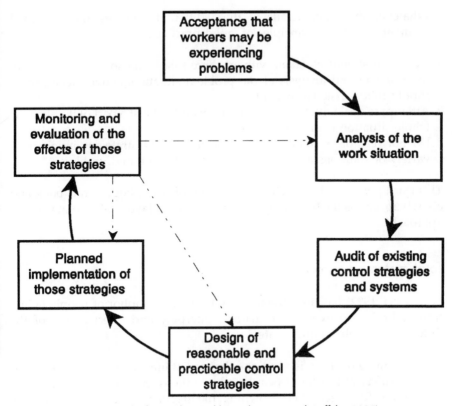

Figure 8.5 The 'control-cycle' for psychosocial hazards (T Cox and Griffiths, 1995)

The United States National Institute of Occupational Safety and Health (Sauter *et al.*, 1990) has proposed a somewhat similar scheme based on four categories of action as part of its national strategy for the protection and promotion of workers' psychological health:

1 job design to improve working conditions;
2 surveillance of psychological disorders and risk factors;
3 information dissemination, education and training;
4 enrichment of psychological health services for workers.

Job design, education and training represent preventive strategies, surveillance may be seen as part of a reactive strategy, while the provision of psychological health services is an important component of any rehabilitative strategy. The most obvious weakness in this scheme (Sauter *et al.*, 1990) exists in relation to 'timely reaction'. There is a need to provide organizational systems and tools to quickly and effectively manage problem situations as they arise and are detected through improved surveillance.

Whatever the exact details of these different schemes, there is good agreement on several principles of 'good practice':

1 No action should be taken without there first being some assessment of the problem and analysis of the problem situation: active management should follow risk assessment.
2 Attempt to deal with the whole situation and not just a part of it unless the whole situation is unmanageable.
3 Do not rely on a single type of solution, build strategies involving a variety of solutions as appropriate to the situation and problem.

This third point has been made in a variety of ways, both in this book and elsewhere, and is the beginning of an argument in favour of a 'total systems approach'.

Total systems approach

Levi (1985, 1992) has used the health education metaphor of people falling from a bridge to explore the nature, interplay and effectiveness of the different available control strategies:

> Spanning the river is a bridge – the road of life – with many defects and no safety rail. A lot of people fall into the river. Many of them cannot swim. To prevent their drowning, the lifeguards in primary healthcare dive into the river, pull them ashore and begin resuscitation. If the lifeguards do not succeed in reaching the drowning people, the latter will fall over a waterfall and sink to the bottom. The divers in our hospitals will then do their best to bring the people to the surface and ashore and subject them to sophisticated and expensive resuscitation.
> It is quite clear that both lifeguards and divers and their institutions and resources are needed. But we also need resources to: repair the bridge, provide the bridge with a safety rail and warning signs, inform people about the dangers of deep water if they cannot swim, teach people to swim and to save other people who cannot swim and need help (Levi, 1985).

An organization-wide or total systems approach to occupational health and safety issues is clearly required. It is neither sufficient nor effective to rely on one type of strategy. In particular, it is not acceptable to simply add in rehabilitative services or facilities, such as work counselling. The whole range of possible strategies needs to be considered and those chosen integrated into a coherent programme. Often this starts with the formulation of an occupational health policy which deals with psychosocial and organizational hazards and their possible health effects, and the specification of the arrangements, resources and organization available to support that policy. Such actions can lead into a careful review of design issues, of management systems and practice, and of the need for enhanced employee support. All actions need to be monitored and evaluated. A common argument against this scale of commitment is cost; however, much can often be achieved within organizations by exploiting existing resources.

Agency and target

Newman and Beehr (1979) have suggested that control strategies for work stress can be categorized according to a three dimensional matrix. Their analysis might be usefully applied here. The first dimension refers to the target (the worker or the organization); the second dimension refers to the type of strategy used, and the third dimension refers to the agency by which the intervention will be accomplished (the organization, external consultants or the workers). This framework can be extended to the present consideration of psychosocial and organizational hazards and reformulated in terms of 'agency–target' pairs. These in turn may be described in the form of three questions which deal with the target of the intervention and control and its agency (T Cox et al., 1990):

1 What can the organization do to put its own house in order (organization:organization)?
2 What can the organization do to help or assist its individual workers (organization:worker)?
3 What can those individual workers do to help themselves (worker:worker)?

The interest here is in the role that the organization can play in managing work-related stress either through 'putting its own house into order' or by enhancing its support for its employees.

What can the organization do to put its own house in order?

The psychosocial characteristics of work which are hazardous have been described in Table 8.1. Health problems may arise because jobs, technology and work environments have not been systematically designed with workers in mind. They might arise because of management style and practice, because of the very culture of the organization, or because change has not been well managed. They may exist because of failures of selection, training, or up-dating workers' knowledge, skills and attitudes. There are therefore three levels at which organizational strategies can operate. They all address issues related to the health of the organization (T Cox and Kuk,

1991; T Cox and Leiter, 1992). The various strategies which might be included here are set out in Figure 8.6.

Given the nature of organizationally focused strategies, there is the question of the role of specialist functions such as occupational health and health and safety. These functions may serve three organizational purposes in addition to their specialist roles: first, they may champion risk

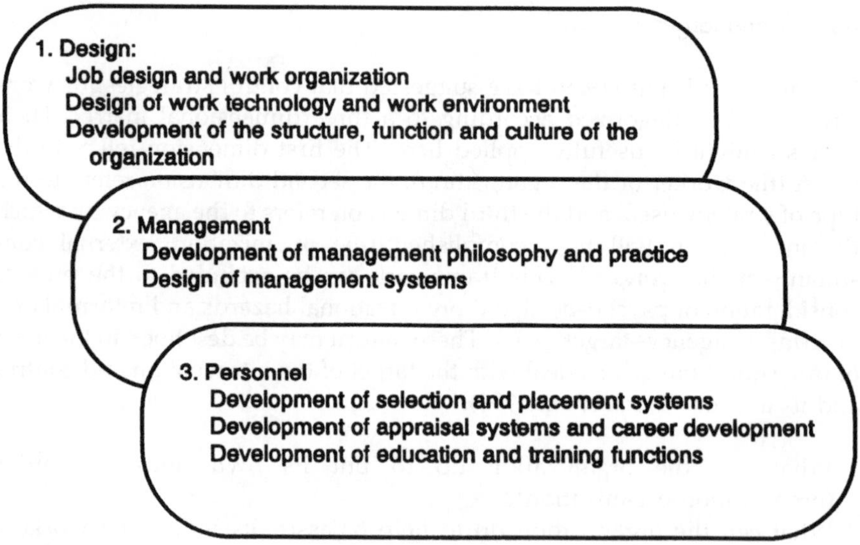

1. Design:
  Job design and work organization
  Design of work technology and work environment
  Development of the structure, function and culture of the
    organization

2. Management
  Development of management philosophy and practice
  Design of management systems

3. Personnel
  Development of selection and placement systems
  Development of appraisal systems and career development
  Development of education and training functions

Figure 8.6 Organizational stress management strategies

management and hazard control in relation to psychosocial and organizational hazards; second, they may provide an integrative overview of such problems and their control; and, third, they may provide the necessary expertise to support action by the organization.

What can the organization do for its workers?

In addition to 'putting its own house in order', an organization might consider how it can provide support for its workers in addition to that received through line management. A relatively small number of large organizations support a traditional occupational health service; others buy into local medical expertise or otherwise employ private medical services. In addition, many organizations, both in the USA and in Europe, offer their staff access to special programmes designed to improve their general health and fitness (health promotion in the workplace), and help them cope with the challenge of work (employee assistance programmes, EAPs). The health promotion activities of organizations in Europe has been reviewed for the WHO Regional Office for Europe by Malzon and Lindsay (1992). The authors of this report were much encouraged both by the number of organizations supporting health promotion programmes and by the number planning to continue them.

In practice health promotion is largely preventive in nature, while employee assistance programmes are largely rehabilitative. However, it can be argued that employee assistance may also fulfil a preventive function, and health promotion may, by the nature of its activities, also be rehabilitative. It is not surprising, therefore, that such programmes are currently converging both in their design and implementation. The evidence suggests that combined programmes have several common elements: the provision of health promotion information (usually smoking cessation, weight control, controlled drinking and diet); fitness and relaxation training, group discussions and/or access to a professional counsellor or better still a personal consultant psychologist; and training in coping skills (such as time management or assertiveness).

Although set up and sponsored by the organization, these programmes can only succeed if the individuals involved are convinced of their value and are drawn into participation. They have to accept at least part ownership of their problems. Much of what is on offer can be taken on-board by those individuals outside of work, perhaps as part of developing a healthier and more robust lifestyle. Thus the question what can the organization do for the individual worker becomes what can the individual worker do for him- or herself?

Effectiveness of employee support programmes

While the literature describing the nature and implementation of different health promotion and employee assistance programmes is substantial, that on their effectiveness is less so. However, attempts at systematic evaluation have been made (Allinson et al., 1989; Hovarth and Frantik, 1991). The results of these studies point up the context dependency of programme effectiveness and the complexity of such initiatives. A useful summary of recent workplace interventions in relation to stress, including employee assistance, has been presented by the ILO (ILO, 1992).

Murphy (1984) concluded that a number of significant benefits accrued to individuals including reductions in physiological arousal levels, in tension and anxiety, in sleep disturbances and in somatic complaints. Some participants have also reported an increased ability to cope with work and home problems following completion of their programme. However, it has been variously noted that many of these benefits rely solely on self-report and that there has been a relative paucity of more objective data in evaluation studies.

Summary

Work-related stress is a health and safety issue, and its experience may result from a breakdown or failure of the work system. This might be a failure of control by design or of control by management. The evidence suggests that the experience of stress can affect the peoples' availability for work and behaviour at work, the quality of their working life and their health. It has been argued in Chapter 5 that such effects may, in turn, contribute to the aetiology of unsafe behaviour at work. As a health and safety issue, work-related stress should be controlled through the application of the risk assessment–risk management paradigm. It should also be considered in all determinations of human reliability probabilities (see Chapter 6 for further details).

References

Ahlbom, A., Karasek, R.A. and Theorell, T. (1977). 'Psychosocial occupational demands and risk for cardiovascular death'. *Lakartidningen*, **77**, 4243–4245.

Allinson, T., Cooper, C.L. and Reynolds, P. (1989). 'Stress counselling in the workplace'. *Psychologist*, September, 384–388.

Cannon, W.B. (1931). *The Wisdom of the Body*. New York: Norton.

Cooper, C.L. and Marshall, J. (1976). 'Occupational sources of stress: a review of the literature relating to coronary heart disease and mental ill health'. *Journal of Occupational Psychology*, **49**, 11–28.

Cox, T. (1978). *Stress*. London: Macmillan.

Cox, T. (1985). 'The nature and measurement of stress'. *Ergonomics*, **28**, 1155–1163.

Cox, T. (1990). 'The recognition and measurement of stress: conceptual and methodological issues'. In *Evaluation of Human Work* (E.N. Corlett and J. Wilson, eds), London: Taylor and Francis.

Cox, T. (1993). *Stress Research and Stress Management: Putting Theory to Work*. Sudbury, Suffolk: HSE Books.

Cox, T. and Cox, S. (1985). 'The role of the adrenals in the psychophysiology of stress'. In *Current Issues in Clinical Psychology* (E. Karas, ed), London: Plenum Press.

Cox, T. and Cox, S. (1993). *Psychosocial and Organizational Hazards: monitoring and control. European Series in Occupational Health no. 5*. Copenhagen, Denmark: World Health Organization (Regional Office for Europe).

Cox, T. and Ferguson, E. (1991). 'Individual differences, stress and coping'. In *Personality and Stress: Individual Differences in the Stress Process* (C.L. Cooper and R. Payne, eds), Chichester: Wiley & Sons.

Cox, T. and Griffiths, A.J. (1995). 'The nature and measurement of work stress: theory and practice'. In *The Evaluation of Human Work: A Practical Ergonomics Methodology* (J. Wilson and N. Corlett, eds), London: Taylor and Francis.

Cox, T. and Kuk, G. (1991). 'Healthiness of schools as organizations: teacher stress and health'. Paper to *International Congress, Stress, Anxiety and Emotional Disorders*, Braga, Portugal: University of Minho.

Cox, T. and Leiter, M. (1992). 'The health of healthcare organizations'. *Work and Stress*, **6**, 219–227.

Cox, T. and Mackay, C.J. (1981). 'A transactional approach to occupational stress'. In *Stress, Work Design and Productivity* (N. Corlett and P. Richardson, eds), Chichester: Wiley & Sons.

Cox, T. and Mackay, C.J. (1985). 'The measurement of self-reported stress and arousal'. *British Journal of Psychology*, **76**, 183–186.

Cox, T., Cox, S. and Thirlaway, M. (1983). 'The psychological and physiological response to stress'. In *Physiological Correlates of Human Behaviour* (A. Gale and J.A. Edwards, eds), London: Academic Press.

Cox, T., Griffiths, A.J. and Cox, S. (in press). *Work-Related Stress in Nursing: Controlling the Risk to Health*. Geneva: International Labour Organisation.

Cox, T. Leather, P. and Cox, S. (1990). 'Stress, health and organizations'. *Occupational Health Review*, **23**, 13–18.

Elo, A-L. (1986). *Assessment of Psychic Stress Factors at Work*. Helsinki: Institute of Occupational Health.

Friedman, M. and Rosenman, R.H. (1974). *Type A: Your Behaviour and Your Heart*. New York: Knoft.

Hovarth, M. and Frantik, E. (1991). 'Mental work stress and health promotion: 15 years follow up'. *Proceedings XXII International Symposium on Behavioural Sciences: Their Role in Health Science and Policy*. July, Tokyo: University of Tokyo.

Hurrell, J.J. and McLaney, M.A. (1989). 'Control, job demands and job satisfaction'. In *Job Control and Worker Health* (S.L. Sauter, J.J. Hurrell and C.L. Cooper, eds), Chichester: Wiley & Sons.

International Labour Office (ILO) (1986). *Psychosocial Factors at Work: Recognition and Control. Occupational Safety and Health Series no. 56*. Geneva: International Labour Office.

International Labour Office (1992). Preventing Stress at Work. *Conditions of Work Digest*, **11**, Geneva: International Labour Office.

Kagan, A. (1975). 'Epidemiology, disease and emotion'. In *Emotions: Their Parameters and Measurement* (L. Levi, ed). New York: Raven Press.

Karasek, R.A. (1979). 'Job demands, job decision latitude and mental strain: implications for job redesign'. *Administrative Science Quarterly*, **24**, 285–308.

Karasek, R.A. (1981). 'Job socialisation and job strain: the implications of two psychosocial mechanisms for job design'. In *Working Life: A Social Science Contribution to Work Reform* (B. Gardell and G. Johansson, eds), Chichester: Wiley & Sons.

Karasek, R. and Theorell, T. (1990). *Healthy Work: Stress, Productivity and the Reconstruction of Working Life*. New York: Basic Books.

Karasek, R.A., Baker, D., Marxer, F. *et al.* (1981). 'Job decision latitude, job demands and cardiovascular disease'. *American Journal of Public Health*, **71**, 694–705.

Kasl, S.V. (1989). 'An epidemiological perspective on the role of control in health'. In *Job Control and Worker Health* (S.L. Sauter, J.J. Hurrell and C.L. Cooper, eds), New York: Wiley.

Kobasa, S. (1979). 'Stressful life events, personality, and health: an inquiry into hardiness'. *Journal of Personality and Social Psychology*, **37**, 1–13.

Kobasa, S. and Puccetti, M. (1983). 'Personality and social resources in stress resistance'. *Journal of Personality and Social Psychology*, **45**, 839–850.

Kobasa, S., Maddi, S. and Courington, S. (1981). 'Personality and constitution as mediators in the stress – illness relationship'. *Journal of Health and Social Behaviour,* **22**, 368–378.

Kobasa, S., Maddi, S. and Kahn, S., (1982). 'Hardiness and health: a prospective study'. *Journal of Personality and Social Psychology,* **42**, 168–177.

Lazarus, R. and Folkman, S. (1984). *Stress, Appraisal and Coping.* Springer Publishing, New York.

Levi, L. (1984). *Stress in Industry: Causes, Effects and Prevention. Occupational Safety and Health Series no. 51.* Geneva: International Labour Office.

Levi, L. (1985). 'Stress: definitions, concepts and significance'. *Cardiovascular Information,* **1**, 10–20.

Levi, L. (1992). 'Psychosocial, occupational, environmental and health concepts, research results and applications'. In *Work and Well Being: An Agenda for the 1990s* (G.P. Keita and S.L. Sauter, eds), Washington DC: American Psychological Association.

Levi, L., Frankenhauser, M. and Gardell, B. (1986). 'The characteristics of the workplace and the nature of its social demands'. In *Occupational Stress, Health and Performance at Work* (S. Wolf and A.J. Finestone, eds), Littleton, MA: PSG Pub. Co. Inc.

Mackay, C., Cox, T., Burrows, G. and Lazzerini, T. (1978). 'An inventory for the measurement of self-reported stress and arousal'. *British Journal of Social and Clinical Psychology,* **17**, 283–284.

Malzon, R.A. and Lindsay, G.B. (1992). *Health Promotion at the Worksite. European Occupational Health Series, No.4.* Copenhagen: WHO Regional Office for Europe.

Murphy, L. (1984). 'Occupational stress management: a review and appraisal'. *Journal of Occupational Psychology,* **57**, 1–15.

National Institute of Occupational Safety and Health (NIOSH) (1988). *Psychosocial Occupational Health.* Washington, DC: National Institute of Occupational Safety and Health.

Newman, J.E. and Beehr, T.A. (1979). 'Personal and organizational strategies for handling job stress: a review of research and opinion'. *Personnel Psychology,* **32**, 1–43.

Payne, R. (1979). 'Developments, supports, constraints and psychological health'. In *Response to Stress: Occupational Aspects* (C. Mackay and T. Cox, eds), Guildford: IPC Science and Technology Press.

Payne, R. and Fletcher, B. (1983). 'Job demands, supports and constraints as predictors of psychological strain among school teachers'. *Journal of Vocational Behaviour,* **22**, 136–147.

Perrewe, P. and Ganster, D.C. (1989). 'The impact of job demands and behavioural control on experienced job stress'. *Journal of Organizational Behaviour,* **10**, 136–147.

Sauter, S.L., Murphy, L.R. and Hurrell, J.J. (1990). 'Prevention of work-related psychological disorders: a national strategy proposed by the National Institute for Occupational Safety and Health (NIOSH)'. *American Psychologist,* **45**, 146–158.

Selye, H. (1950). *Stress.* Montreal: Acta Incorporated.

Selye, H. (1956). *Stress of Life.* New York: McGraw-Hill.

Spector, P.E. (1987). 'Interactive effects of perceived control and job stressors on affective reactions and health outcomes for clerical workers'. *Work and Stress,* **1**, 155–162.

Szabo, S., Maull, E.A and Pirie, J. (1983). 'Occupational stress: understanding, recognition and prevention'. *Experientia*, **39**, 1057–1180.

Warr, P.B. (1990). 'Decision latitude, job demands and employee well-being'. *Work and Stress*, **4**, 285–294.

Warr, P.B. (1992). 'Job features and excessive stress'. In *Prevention of Mental Health at Work* (R. Jenkins and N. Coney, eds), London: HMSO.

Part 3
Safety Management

The authors have set out their framework for the management of safety in the preceding eight chapters (Parts 1 and 2). They have described how it is built on systems theory, and considers the individual in the job (tasks) in the organization. The individual, in this context, is treated as an active information processor. The preceding eight chapters have set out the theory; the following four chapters (Part 3) explore the implications of the theory for practice.

Part 3 is concerned with two fundamental objectives: first, the design of safe and healthy organizations and safe systems of work, and second, the effective management of work and safety. The first objective relates to control of safety by design, while the second relates to the control of safety through ongoing management. The former has been described in Chapter 4 as non-adaptive control, and the latter as adaptive control. All four chapters in Part 3 contain discussions of both. The four chapters are:

9 Safety: management responsibilities and practices;
10 Safety management: strategies;
11 Managing the work environment: the design of safe work;
12 Managing people and their attitudes to safety.

Although each of these four areas is the subject of a separate chapter, the conceptual and operational overlaps between them should be obvious. The authors firmly believe that the effective management of safety is dependent on their interdependence being recognized, taken into account and exploited. Consider the following case study.

A company involved in the manufacture and storage of chemicals had experienced a series of serious accidents at one of their plants, including a fatal accident to a contractor, in a relatively short period of time. Concerned to improve the company's safety performance, senior management resolved to make immediate improvements. A cursory, rather than a systematic, study of each accident lead those senior managers to believe that it had been the unsafe behaviour and negative attitudes of operational staff which had led to the series of accidents. The operational staff group was blamed for the company's poor safety record. A behavioural safety programme focused on that group was 'bought in', and, although small improvements were achieved in the first six months, these were not sustained in the longer term. Furthermore, the attribution of blame implicit in the actions taken by senior management had caused a deterioration in staff morale. A second and, this time, systematic study of the company's safety performance was commissioned from an external agency. It was quickly established that the company's problems were related to a lack of management commitment to safety and poorly designed systems of work *combined* with poor safety engineering controls and unsafe behaviour on the part of employees. While those employees' behaviour was implicated in the aetiology of the company's accidents, that behaviour was just one part of a wider problem scenario involving not only the commitment and behaviour of management but also aspects of the organization's safety hardware and software (systems and procedures). An approach which integrated attacks on all these various and different aspects of poor safety performance seemed to be the company's only way forward.

This message is the practical backbone of this text. It has been repeated many times in the first eight chapters, and will be repeated again in the following four chapters.

9
Safety: management responsibilities and practices

Introduction

The term 'management' is used in a number of different, but obviously related, ways. It is used to refer to an organizational *function*, as in the management of safety. The management of safety, as a function, embraces both control by design and control by the ongoing process of managing. The term is also used to refer to that *process*, managing, and, finally, as the designator of a level or *group* within an organization – the management. Much of this text has concerned the management of safety as an organizational function and discussed a framework for thinking through that function. This chapter focuses much more on managers as a group and explores their responsibilities and the requirements and strategies for managing that are associated with those responsibilities.

Chapter 9 is a 'watershed' within the book; it marks a transition from discussions of theoretical frameworks for the management of safety to those related to the practical aspects of managing. It is primarily concerned with management actions, and considers:

1 the primary tasks and responsibilities that managers in all organizations have for managing safety;
2 the practices that support compliance (including current regulatory guidance); and finally
3 the importance of management input into organizational development for safety.

It also considers, in broad terms, the legal duties held by managers with respect to safety management. It does not explore those duties in detail; this is not necessary at this stage and is beyond the scope of the book.

Drucker in his book on *Management: Tasks, Responsibilities and Practices* (Drucker, 1994) has defined the three primary tasks (essential to all, but organization specific) that face the management of every organization. These include:

1 thinking through and defining the specific policies and goals of the organization (goal-setting);
2 making work 'productive' and enabling the worker to 'achieve'; and
3 managing social impacts and social responsibilities.

Drucker (1994) deems these tasks to be the three *dimensions* of management. These dimensions can be applied to safety and are discussed below.

Goal setting

All organizations exist for a specific purpose. For example, a university exists to further the education of its students and a business enterprise to make a profit through the sale of its products (or services). Although economic performance is important to both types of institutions, Drucker (1994) argues that only business has economic performance as its *central* goal. Despite this, recent initiatives in safety management have recognized the importance of economic arguments for all organizations (see, for example, *Safety Pays – Six Business Reasons for Managing Health and Safety at Work*, an initiative sponsored by The Engineering Employers' Federation (EEF) and Lloyds Bank, 1995). Slogans such as 'Safety is good business' (HSE, 1993) also reflect this move. The argument is that the effective management of safety can be a way of achieving central goals (quality and production) and is thus an integral part of the process underpinning the primary purpose. Organizations have a number of different goals and objectives. Simon (1981) showed that management within successful organizations did not optimize on one goal or objective at the expense of all others (see Chapter 4). What they did was to find 'good enough' solutions across their range of goals and objectives, trying to balance one against another. This is the principle of satisficing. It has implications for the importance assigned to safety management, and its goals and objectives. The arguments have been expressed in policy statements such as: 'nothing is more important than safety, not production nor profits' (Griffiths, 1986); and 'nothing is so important that we cannot take the time to do it safely'.

The strategic approach to the management of safety demands two complementary processes in this domain (see Figure 9.1):

1 First – management has to decide where safety fits in relation to the primary purpose and goals of their organization (this is represented in Figure 9.1 as a 'Level 1' decision).
2 Second – they have to decide on the goals and objectives for safety management. At this stage they may also consider how safety goals can be integrated into all areas of business activity.

Managers within organizations such as Air Products (Griffiths, 1986; S Cox, 1988) have elaborated their thinking in relation to the first step, which designated safety as having an equal priority to all other functions (Griffiths, 1986), to produce goals which strived to make the organization 'the best in the business' in safety performance. Managers in many other organizations have tempered this objective and chosen to strive for

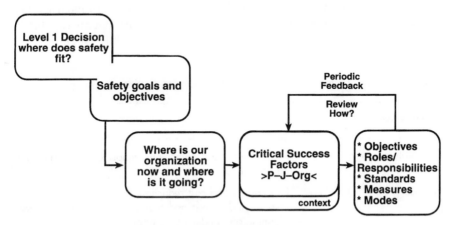

Figure 9.1 Organizational planning for 'safety': mission and purpose

continuing improvements in safety, in line with their approach to quality. So what should the goals of safety management be and how should individual managers approach the task of defining them?

Goals of safety management

The goals of safety management are to manage safety effectively and in doing so to intervene in the accident process such that accidents and incidents (and the severity of any such events) are minimized (Advisory Committee for Safety in Nuclear Installations (ACSNI), 1993). The primary goal is good safety performance. The processes involved should be active, rather than passive, and it is recommended that they are set in the context of the 'risk assessment – risk management' paradigm (see Chapter 2). They should involve the following sequence of actions:

1 effective hazard identification;
2 suitable and sufficient assessment of risks;
3 effective control procedures and mechanisms;

4 monitoring; and
5 evaluation of existing processes and rewards.

It has already been argued that the goals of safety management should not simply be limited to the safety domain but should be integrated into all areas of business activity. Indeed, research (HSE, 1981, 1991) has shown that the management systems and characteristics required to achieve success in safety are exactly the same as those required for other areas of organizational activity for example, production and quality. Those organizations that are capable of managing safety also perform well in the other domains (HSE, 1981). Table 9.1

Table 9.1 Impact of effective health and safety policies on 'business' thinking

Business 'element'	Areas of impact
Corporate strategy	Mission statement, community image, and management competencies
Finance	Loss control and cost reduction, risk management, investment decisions and property acquisition, financial planning and budgetary control
Human resources	Recruitment and selection, training and development, culture, work organization, health promotion and stress management
Marketing, service specification and liability	Specification of service, and health and safety standards
Operating policy	Design of jobs and equipment, maintenance of premises, environmental management and waste disposal
Information management and systems	Identification of critical data, selection of appropriate performance indicators, utilization of information technology in data collection (etc.)

(adapted from S Cox *et al.*, 1995) highlights the areas of 'business' thinking (in addition to production and quality) which interact with safety management. The information presented in this table may enable managers and policy makers to extend the scope of traditional safety management (see Chapter 3) and help them integrate safety management into mainstream policy decisions. For example, training and development policies which address core competencies should also address those required for safety.

Goal setting and decision making

Decisions on safety goals should be based on a shared vision (see Chapter 10), the common understanding of the organization and an appreciation of the direction and efforts in this area (Drucker, 1994). They require a definition of the following:

1 'where is our organization now and where do we want it to be?'; and
2 'what are the critical success factors in relation to the person-job-organization model?' (outlined in Chapter 4).

These decisions should be conditioned by the nature of the organization and its context; for example, the aspirations of a small- to medium-sized enterprise (SME) operating in a relatively 'risk-free' environment will be different from those of a multinational organization involved in production of oil and petroleum. At the same time, some discussion should be devoted to the manner in which the organization could achieve success (see Figure 9.1) through the use of strategies such as role definition and standard setting (see Chapter 10). It should also consider issues of feedback as part of ongoing control.

Pool and Koopman (1989) have proposed a decision making model which considers several opportunities for choice set in specific organizations (see Figure 9.2). Their model is based both on a review of management decision making theories and their own research in this area (Koopman and Pool, 1994). The factors which appear to impact on the opportunities for choice are: (1) the complexity (see Chapter 10) and defining characteristics of the organization; (2) the personalities of key decision makers; (3) the prevailing physical and social environments; and (4) the perceived importance of safety. The application of this thinking to the formulation of safety goals requires a fuller understanding of four issues.

Centralization

Centralization relates to the extent to which senior management keeps decision making to itself or involves other key groups, parties or hierarchical levels.

Heller et al. (1988) have shown this element to be the primary source of variance in the effectiveness of the decision making process. For 'safety', key groups could include: safety representatives and committees, safety advisers, occupational health professionals, risk assessment coordinators, line managers and supervisors and specific job holders. Aspects of 'centralization' relate to the transparency and management of the consultation process as well as the eventual outcomes. It should be noted that senior management has been criticized in the past for delegating decision making on safety to individuals or groups with a low powerbase (HSE, 1981). They

Figure 9.2 Organizational decision making for 'safety' (adapted from Koopman and Pool, 1994)

have been encouraged to make their own decisions. This should be balanced against the requirements for consultation enshrined in various parts of UK and EU legislation.

Formalization

The formalization dimension relates to the degree to which decision making is formalized.

Decisions can take place according to a formal procedure laid down in advance or they can proceed more flexibly on an ad hoc basis. They may also be attached to the formal consultation and communication structures or be incorporated into temporary structures. Current approaches to safety favour clearly defined communication procedures (HSE, 1991).

Information

The dimension of information comes from classical decision making theory, which is based on the assumption that decision makers not only have access to all the necessary information but are also able to process it (Harrison, 1987). This can be a particularly dangerous assumption in the safety domain.

Information in the field of safety performance has traditionally been retrospective, and related to lost time accidents, incidents and sickness absence. Furthermore, such information has often been incomplete (Davies and Teasdale, 1994). Ideally, decisions should be made on the basis of broad systemic knowledge (see Chapter 10) and on the contributions of external agencies. Dawson (1992) has reviewed the necessary information required for safety decisions including scientific and technical knowledge of hazards and risks, information on management systems including emergency preparedness and employee behaviours. She has particularly stressed the importance of knowledge of human behaviour and organizational processes in addition to that which is derived from natural science and technology.

Conflict and confrontation

The final dimension – conflict and confrontation – relates to the extent to which there are conflicts and thus, by implication, confrontations in the decision making process.

Conflicts have always existed (to a greater or lesser extent) in the safety domain. These conflicts have related to the importance and perceived costs of safety management in relation to other functions. They emerge in decisions at all levels in the organization. They are also evident in the continuing debates on public perceptions of risk (see, for example, Douglas, 1992; Sjöberg, 1995; Slovic, 1993).

It has been argued that consideration of the four issues provides the basis for sound decision making practice (Koopman and Pool, 1994).

Pool and Koopman's (1989) model also refers to environmental characteristics (geography and risk), and political considerations (visibility of safety management and previous incident and accident history). For example, decisions on safety within the nuclear industry are always high profile and any incident which occurs (however minor) attracts vast media attention.

Productive work and worker achievement

Drucker's second task is to make work productive and to facilitate worker achievement (Drucker, 1994). In the context of safety, this task encompasses the realization of both safe and productive work. This can only be achieved, at least in the long term, through the design of healthy organizations.

Healthy organizations, according to the emerging theory (see Chapter 5) are ones in which:

1 the technical and social organizations are congruent;
2 employee perceptions of the organization are realistic;
3 the task, problem solving and development sub-systems are adequate; and
4 these sub-systems are well coordinated through effective communication and a strong social milieu.

These four characteristics of a healthy organization can be easily transformed into practical objectives and may become the focus of organizational development activities (see Chapter 10). The manager's role is to promote these objectives and facilitate their achievement. This can be done in a number of ways, some of which overlap with their other two 'tasks' (Drucker, 1994). At the highest level, they can articulate appropriate goals and influence attitudes and culture. They can also ensure the necessary systems and social interactions, and rewards through their day-to-day activities. Finally, they can initiate additional organizational development activities focused on the theory of organizational health. This is mentioned again in the next chapter (Chapter 10). A common device, and one used by the authors in this context, is the development workshop. This has proved successful in many contexts in promoting the health of organizations, but only when properly structured and executed.

Some of the issues raised in relation to 'healthy' organizations carry over into a discussion of Drucker's (1994) third task of management.

Social impacts and social responsibilities

The third task of management (Drucker, 1994) is managing the social impacts and social responsibilities of the organization. No organization exists by itself or as an end in itself. Each organization has its place in the wider society and social environment and to that extent it reflects the aims and objectives of society as a whole. The controversy surrounding the disposal of an obsolete oil-rig (*Independent* Newspaper, 20 June 1995) illustrates this point.

The UK legislative framework could be deemed to encompass social objectives in safety; thus knowledge and understanding of the relevant legislation, by management, is a necessary prerequisite to performing this third task. To help in acquiring this knowledge, open and honest dialogues with the appropriate enforcement agencies (either the Health and Safety Executive (HSE) inspectorate or the Local Authority environmental health officers) are encouraged.

The legal context

The legislative framework for managing safety in the UK is enshrined in the 1974 *Health and Safety at Work etc. Act* (HASAWA). HASAWA, in contrast to previous attempts to improve occupational safety, sought to establish a new approach (Robens, 1972) by encouraging the philosophy of self-regulation. Active safety management was seen to be central. Subsequent developments within the European Union (EU) have confirmed the importance of such an approach and have formalized the requirement for risk assessment (see Framework Directive 89/391/EEC). These requirements are contained in the recently enacted Management of Health and Safety at Work Regulations 1992 (MHSWR) (HSC, 1992) which were discussed in Chapters 1 and 4. Regulations 3 and 4 of MHSWR require employers to carry out a risk assessment (see later) and to demonstrate evidence of a management system as part of the necessary control systems. Managers are pivotal in this process.

The UK Health and Safety Executive (HSE) have published guidance on the design and implementation of health and safety management systems – *Successful Health and Safety Management* (HSE, 1991). This publication describes the main elements of a safety management system and thus provides a framework which enables managers within organizations to achieve compliance with the MHSWR. The approach favoured by the HSE involves the application of the principles of total quality management (see, for example, Crosby, 1984; Deming, 1986; Juran, 1964) to safety. This approach is consistent with the argument that organizations (and their managers) should approach safety management as a mainstream activity, and treat it in a similar manner to other management issues. It is also consistent with a total loss prevention approach as outlined in Bird and Germain (1986).

The key elements of the safety management system recommended by the UK HSE (1991) are set out in Figure 9.3. They include:

1 developing policy, which sets out the organization's general approach, goals and objectives towards safety issues;
2 organizing, which is the process of establishing the structures, responsibilities and relationships which shape the total work environment;
3 planning, the organizational process which is used to determine the methods by which specific objectives should be set and how resources are allocated (risk assessment is an integral part of the planning process);
4 implementing, which focuses on the practical management actions and the necessary employee behaviours required to achieve success;
5 measuring performance, which incorporates the processes of gathering the necessary information to assess progress towards safety goals; and
6 auditing and reviewing performance, which is the review of all relevant information and the process of reflection. It is the final step in the cycle

which defines a systems approach to safety management and which draws heavily on feedback for effective control.

Figure 9.3 (adapted from *Successful Health and Safety Management*, HSE, 1991) illustrates how feedback on performance may be used in the ongoing review (including various types of audit) and development of each of the key elements. It is part of the system control mechanism described in Chapter 4. The legal requirement for the audit of an organization's safety management system is also enshrined within a number of UK safety regulations (see, for example, The Offshore Installations (Safety Case) Regulations 1992).

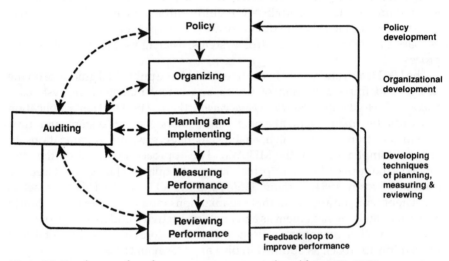

Figure 9.3 Key elements of a safety management system (adapted from HSE, 1991)

Procedures and protocols (S Cox and Fuller, 1995) for safety management system audit and review will be elaborated in Chapter 10 and are illustrated in an appendix to this text. The following section considers the elements presented in Figure 9.3.

Policy

Policy is concerned (in the broadest sense) with the way in which the organization orientates itself to challenges and opportunities in the safety domain. It often describes the principles underpinning its behaviour. For example, the organization may decide to make excellence in safety the thrust of its marketing policy (Confederation of British Industry (CBI), 1990). Equally, it may decide to invest in advanced driver training in order to increase the driving skills of at-risk personnel as part of its insurance policy. Policy formulation is one of the three primary tasks of management which have been described by Drucker (1994).

This active and dynamic process of policy development should be distinguished from the legal requirement to produce a written statement of health and safety policy (commonly referred to as the 'Company Safety Policy') required under Section 2 of HASAWA (1974). The distinction is of particular importance when the research on effectiveness of safety policies is considered (APAU, 1982). Studies carried out by the Accident Prevention Advisory Unit (APAU, 1982) of the HSE into the development and use of written safety policy documents showed their effectiveness to be generally poor. In many cases policies did not even meet the minimal legal requirements. Furthermore, the duties of operational staff were more fully articulated than those of managers within these policies. The emphasis was on the duties of 'safe workers' rather than on management roles and control systems.

Managers' roles
In order to fully secure such policies HSE has recommended in their publication, *Successful Health and Safety Management* (HSE, 1991), that managers:

1 develop appropriate organizational structures and a culture which supports risk control (see Chapter 10);
2 adopt a total loss approach (Bird and Germain, 1986) and link the prevention of harm to employees to the preservation of physical assets;
3 stress the importance of management control and accept the fact that the majority of accidents and incidents are not caused by 'careless workers' (Lee, 1993; Reason, 1990) but by failures in control (or organization of the job); .
4 take a systematic approach (based on risk assessment (S Cox and Tait, 1991)) which adopts logical business principles; and
5 establish the links with total quality management (see later).

The messages conveyed by effective health and safety policies thus pervade all areas of business activity.

Organizing

Organizing is the process of designing and establishing the structures, responsibilities and relationships which shape the total work environment in order to secure the implementation and development of safety policy. The HSE (1991) describe the four Cs of organizing: control, cooperation, communication and competence.

1 *Control* is the foundation of a safety culture (Alexander *et al.*, 1994). It is achieved by securing the commitment of all employees to safety objectives. All line managers should have safety responsibilities allocated to them and these should be formulated in a manner which is consistent with defined responsibilities in other domains. A senior figure at the top of the organization should be nominated to coordinate and monitor policy implementation.

2 *Cooperation* between employees and their representatives (see Chapter 10) is an essential ingredient of an effective safety system. There are several ways to secure employee involvement in safety matters including:
   (a) formal safety representation and committees;
   (b) the implementation of safety suggestion schemes;
   (c) the active encouragement of safety circles (similar to quality circles), where safety problems are identified and solved;
   (d) the use of safety action and discussion groups (Kjellén and Baneryd, 1983) who focus on a specific safety project within the organization; and
   (e) the use of 'near miss' or hazard report forms in support of organizational learning (van der Schaaf *et al.*, 1991).

3 *Communication* is involved in all aspects of work and is a vital process in all organizations. Communication processes and networks have been variously studied (see, for example, Levitt, 1951). For safety they are as much concerned with information coming in and going out of the organization, as they are with ensuring adequate transfer and flow within. Sources for securing the appropriate 'safety intelligence' coming into the organization are essential. There are various ways in which an organization may

Table 9.2 Organizational communication: routes and processes

Generalized organizational communication systems	• management and team briefings • company newsletter and magazine • notice-boards, posters, etc. • suggestion schemes
Health and safety communications	• the organization's health and safety policy • safety training • safety committees and representatives • campaigns, slogans, aides-mémoire, etc. • written safety instructions • the distribution of HSE leaflets, etc.
Management and employee processes	• day-to-day management and supervisory practices • 'setting an example': management and supervisory behaviour • peer interactions

communicate safety issues. Table 9.2 groups these into three areas: generalized organizational communication systems, specific safety communications and management, and employee communication processes.

4 *Competence* is a vital ingredient of safety behaviour. It relates firstly to the necessary skills for successful task completion (S Cox *et al.*, 1995; Saunders and Wheeler, 1991) and secondly to the safety-specific skills (S Cox, 1987; Dawson *et al.*, 1988). The necessary competencies can be secured through training (see Chapter 12).

Planning and implementing: risk assessment

Planning and implementation are the two elements at the heart of successful management (see Figure 9.3) and together they form the basis of risk assessment which is now explicitly required by the safety legislation throughout the EU (Framework Directive 89/391/EEC).

The legal requirement for a risk assessment of safety-related workplace hazards is not a new concept. Legislation governing the use of asbestos

Table 9.3 Key elements of RMS (Risk Management Software) (Warwick IC Systems and CHaRM, 1995)

Element 1	**Risk assessment** Hazard codes and a free text description form the basis of the assessment. Hazard codes can be 'tagged' with existing controls, numbers at risk and quantified in terms of likelihood and consequences via a user-definable risk rating matrix.
Element 2	**Actions/controls** Actions and controls, both standard and specific for the hazard identified within the risk assessment, can be recorded. The attached priority, key person responsibility and status code, help to ensure that the ongoing requirements can be managed effectively.
Element 3	**Personnel** The personnel module incorporates records of employment history, training and document issue to provide effective control of employee competence and exposure to defined work environments. 'At risk groups' can be linked to a risk assessment to highlight training and health surveillance requirements based on the identified hazards.
Element 4	**Accidents** Details of an incident can be recorded via a series of user-definable codes to allow for effective and structured analysis. Key facilities include details of injuries, causal factors, remedial actions and associated cost implications.

(HMSO, 1987), hazardous substances (HMSO, 1988), and noise (HMSO, 1989) all impose duties on the employer to assess the risks from specific workplace hazards and then to identify and implement appropriate measures to reduce those risks. Regulation 3 of MHSWR (see Chapter 2) requires every employer and self-employed person to carry out a 'suitable and sufficient' risk assessment and the results should be recorded (either in writing or electronically) if the organization employs more than five persons.

The first author (S Cox, 1992; Walker and S Cox, 1995) has developed risk assessment procedures (both paper based and electronic) for compliance with Regulation 3 of MHSWR. The second author and his colleagues at Nottingham have somewhat similarly developed risk assessment procedures for psychosocial hazards and work-related stress (T Cox *et al.*, 1995). Table 9.3 outlines the key elements of a computer-based system (RMS) developed by Warwick I.C. Systems in collaboration with Walker and S Cox (1995).

The processes underlying an effective assessment of workplace risks have been described in Chapter 2. What is stressed here are the legal considerations, together with the need for up-to-date and user friendly administrative systems (see Table 9.3).

Performance measurement

Measurement is an essential aspect of maintaining and improving safety performance. It enables managers to ensure that standards achieved in practice are in line with the key objectives (outlined earlier).

HSE (1991) elaborate two key requirements for measurement in the safety area:

1 proactive systems which *monitor* the achievement of plans and the extent of compliance with standards; and
2 reactive systems which *monitor* accidents, occupational ill-health and near miss incidents.

Both systems should generate information on levels of performance if they are supported by the appropriate data recording systems. Performance measurement is further discussed in the next chapter of the book.

Auditing

Auditing is an essential process in the safety management cycle and it provides managers with further information on compliance with standards. Auditing can be distinguished from inspection and routine monitoring in that it provides an objective and formally documented overview of the whole management system (T Cox and S Cox, 1993). Audit systems can be devised by external agencies and there are many proprietary systems available. Alternatively, companies can devise their own systems.

The audit process may be initiated and carried out within the organization, provided that competent auditors are used and that they are auditing areas outside their immediate line control and responsibility. Equally, the audit may be carried out by independent auditors. It is for each undertaking to determine the scheme which is best suited to its needs and, in the case of proprietary systems, this often involves choices on costs and practical benefits. Further information on auditing is outlined in the next chapter.

Performance review

The process of safety management described in this chapter is similar to all management by objective systems (Waring, 1989). These systems are all characterized by the need for ongoing review. Indeed, if the review process is curtailed then continuous improvement is impossible. Systems should be in place to secure an effective review. The system should specify the period over which the review takes place, the nature of the performance data and the reviewers. In practice, the review should be undertaken by the senior management and any resulting change in policy should be agreed with representatives of the workforce.

The safety management system described in Figure 9.3 is proactive and purposeful. Managers who successfully adopt this approach not only secure compliance with the MHSWR (HSC, 1992) for their organization, but also support quality improvements.

Safety management and quality

An increasing number of managers are recognizing the similarities between safety and quality. Whiston and Eddershaw (1989) have applied the Crosby (1984) 'absolutes' of quality to safety. The four 'absolutes' include:

1 definition – quality is conformance to agreed standards;
2 system of quality – prevention focused on work processes;
3 standards of quality – zero defects;
4 measurement of quality – the cost of non-conformance.

Table 9.4 (adapted from Whiston and Eddershaw, 1989) illustrates how Crosby's (1984) principles can provide a framework for safety considerations. Together these four principles parallel some of the elements of the safety management system (HSE, 1991), in particular the elements of policy, organizing, planning and performance monitoring. Auditing against clearly defined standards is also central to the Crosby (1984) approach.

The fact that performance measurement indicators are expressed in monetary terms is a clear advantage of the application of the Whiston and

Table 9.4 The four Crosby principles applied to safety (adapted from Whiston and Eddershaw, 1989)

Conformance to agreed requirements
Job/task requirements are clearly established, agreed and owned by all employers. Written and auditable procedures define how to do the job properly and safely. Sufficient resource – equipment, people and training – is deployed by management to ensure full conformance with safe working.

Prevention focused on work processes
Plans are developed and implemented to improve safety performance through policies and systems for:
● education of all employees;
● communication of standards;
● open investigation of accidents and near misses aimed at identifying root causes and preventing recurrence;
● regular auditing to identify potential sources of error.
No work is done without prior consideration of safety standards and of possibility for errors and error avoidance.

Performance standards
Everyone believes and shows by personal neverending effort for improvement that it is possible to prevent accidents. Accidents are regarded as a personal affront and an opportunity to learn (some organizations go further and embrace the philosophy of zero accidents).

Measurement – the cost of non-conformance
Actual non-conformances such as accidents and incidents are expressed in financial terms, as are the potential non-conformances or near misses. These measurements are used for eliminative or corrective actions.

Eddershaw framework. However, this advantage may only be realized in practice if management are made accountable for losses in their own area. It is also dependent on the organization's overall approach to quality and its integration with safety systems.

A survey of 300 engineering companies (reported by Walter, 1995) was carried out as part of an ongoing study on the relevance of reported quality initiatives to health and safety performance within the Engineering Employers' Federation (EEF). It was found that, of the sixty-nine companies that responded to the survey:

1 fifty-three companies (77 per cent of respondents) had implemented a quality improvement programme; however, only six (9 per cent) had fully and twenty-four (35 per cent) had partially integrated health and safety within this programme;

2 twenty-three (33 per cent) had not integrated health and safety into their quality management programmes whatsoever, even though many of these programmes were long established (up to five years in some cases);

3 the most common type of quality improvement programme was British Standard BS 5750 (BSI, 1987); and

4 thirty-two (46 per cent) of respondents believed that there was a need for a British Standard (BS) for health and safety management systems similar to BS 5750. However, an almost equal number, thirty-one (44 per cent), did not subscribe to this view.

Although these findings only reflect the views of a relatively small sample of companies within the engineering sector, there is some indication that safety is beginning to become integrated into other areas of business activity. The views expressed on the need for a BS on the management of safety are equivocal and possibly reflect the relative newness of management systems' approaches in the safety arena.

Standards for safety management systems

The draft BS (at the time of writing this book), Guide to Safety Management Systems, is based around the elements identified by the HSE in their publication (described in Figure 9.3). However, there is an additional element in the BSI approach (BSI, 1995) – an 'initial and periodic status review'. Such a review provides managers with a measure of their current performance in safety and is an integral part of the authors' approach (see Chapter 10). The BSI model also adopts an integrated systems approach (see Figure 9.3) in which the safety management system elements are incorporated into the wider organizational system, which is in turn moderated by external factors (see Chapter 4). It should be noted that this draft BS does not in itself lay down safety performance criteria, nor does it seek to give detailed guidance on safety management system design. It only provides guidelines and a further framework for action and as such reinforces the approach described earlier (HSE, 1991). Similar models have been described by the professional institutions, for example the Institution of Chemical Engineers (IChemE, 1994) in a publication compiled from contributions of technical representatives from the European Process Safety Centre (EPSC) Working Party on Safety Management Systems.

Generic safety management model

Each of the primary tasks (Drucker, 1994) described in the previous sections have obvious implications in the safety domain. The final section of this chapter presents a generic safety management model (see Figure 9.4) as a

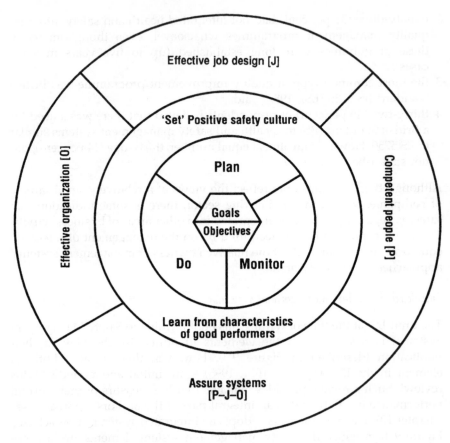

Figure 9.4 Generic safety management model

framework for the following three chapters. The model seeks to further integrate the safety goals and approaches discussed earlier in this chapter with the systems framework outlined in Chapter 4.

It brings together:

1 basic systems concepts of control and particularly purposeful control through planning and effective decision making (Koopman and Pool, 1994) and monitoring;
2 goal setting and its centrality in current approaches to safety legislation;
3 effective task implementation and completion (Drucker, 1994);
4 the need to establish a 'good' safety culture which incorporates the characteristics of good performers outlined in Chapter 5 (Lee, 1993, 1995);
5 the implementation of a Person–Job–Organization approach which requires the organization to: (a) 'set' the organization (see Chapter 10); (b)

define technical standards and safe systems of work and (c) define competencies for its employees (see later); and

6 finally, the need to assure all P–J–O systems in line with quality principles (see Chapter 10).

The model could be likened to an onion. As the onion grows, each layer builds upon the next until finally it is matured and fully grown. The essential core is the setting of goals and objectives (the definition of purpose) and the utilization of the 'plan-do-monitor' approach. The next layer recognizes and represents the importance of safety culture and the outer layer is the effective manifestation of the system in person–job–organization terms. It is the manifestation of a successful safety management system and provides a blueprint for effective design.

Summary

This chapter has attempted to explicate the responsibilities and requirements on managers in relation to safety. In doing so, it has explored the legal imperative in the design of the safety management system. It has also highlighted the similarities between safety and quality. Throughout, Drucker's (1994) three tasks of management have provided a scaffold for the discussion. Finally, a generic management model was proposed and is further developed in the following chapters. Chapter 10 develops the organizational dimensions in Figure 9.4 and considers the concept of safety culture. Chapter 11 focuses on the design of 'safe' jobs and the final chapter considers issues relating to the individual within the system.

References

Accident Prevention Advisory Unit (APAU) (1982). *The Effectiveness of Safety Policies.* London: HMSO.

Advisory Committee for Safety in Nuclear Installations (ACSNI) (1993). *Organising for Safety – Third Report of the Human Factors Study Group of ACSNI.* Sudbury, Suffolk: HSE Books.

Alexander, M., Cox, S.J. and Cheyne, A. (1994). 'Safety culture within a UK organisation engaged in offshore hydrocarbon processing'. In *Proceedings of the IVth Conference on Safety and Well-Being at Work, November 1–24,* pp. 64–74. Loughborough: Loughborough University of Technology.

Bird, F.E. and Germain, G.L. (1986). *Practical Loss Control Leadership.* Loganville, GA: Institute Publishing.

(British Standards Institution) BSI (1987). British Standard 5750. *Quality Systems: Specification for Design, Manufacture and Installation.* London: British Standards Institution.

(British Standards Policy Committee OHS) BSI (1995). *Draft Guide to Health and Safety Management Systems, Document 941 408875*. London: British Standards Institution.

Confederation of British Industry (CBI) (1990). *Developing a Safety Culture – Business for Safety*. London: Confederation of British Industry.

Control of Asbestos at Work Regulations (1987). *Statutory Instrument 1987 No. 2115*. London: HMSO.

Control of Substances Hazardous to Health Regulations (1994). *Statutory Instrument 1994 No. 3246*. London: HMSO.

Cox, S.J. (1987). 'Safety training: an overview of current needs'. *Work and Stress*, **1**, 67–71.

Cox, S.J. (1988). *Employee Attitudes to Safety*. Unpublished M.Phil. thesis, University of Nottingham.

Cox, S.J. (1992). *Risk Assessment Toolkit*. Loughborough: Centre for Extension Studies, Loughborough University of Technology.

Cox, S.J. and Fuller, C. (1995). *Safety Managment Systems Audit Protocol*. Loughborough: Centre for Hazard and Risk Management, Loughborough University of Technology.

Cox, S.J. and Tait, N.R.S. (1991). *Reliability, Safety and Risk Management*: An integrated approach. Oxford: Butterworth Heinemann.

Cox, S.J., Janes, W., Walker, D. and Wenham, D. (1995). *Office Health and Safety Handbook*. London: Tolley.

Cox, T. and Cox, S. (1993). *Psychosocial and Organisational Hazards: Monitoring and Control. European Series in Occupational Health no. 5*. Copenhagen, Denmark: World Health Organization (Regional Office for Europe).

Cox, T., Griffiths, A. and Cox, S.J. (1995). *Work-Related Stress in Nursing: Managing the Risk*. Geneva: International Labour Office.

Crosby, P.B. (1984). *Quality Without Tears*. New York: McGraw Hill.

Davies, N.V. and Teasdale, P. (1994). *The Costs to the British Economy of Work Accidents and Work-Related Ill-Health*. Sudbury, Suffolk: HSE Books.

Dawson, S.J.N. (1992). 'Knowledge is not enough: developing the capacity for self regulation at work'. In *Managing Organisations in 1992: Strategic Response* (P. Barrar and C.C. Cooper, eds), London: Routledge.

Dawson, S., Willman, P., Clinton, A. and Bamford, M. (1988). *Safety at Work: The Limits of Self Regulation*. Cambridge: Cambridge University Press.

Deming, W.E. (1986). *Out of the Crisis*. Cambridge, MA: MIT Press.

Douglas, M. (1992). *Risk and Blame*. London: Routledge.

Drucker, P.F. (1994). *Management: Tasks, Responsibilities, Practices*. Oxford: Butterworth Heinemann Ltd.

Engineering Employers' Federation (EEF) and Lloyds Bank Commercial Service (1995). *Safety Pays – Six Business Reasons for Managing Health and Safety at Work*. London: EEF.

European Commission (1989). *Framework Directive on the Introduction of Measures to Encourage Improvements in the Safety and Health of Workers at Work*. 89/391/EEC.

Griffiths, D.K. (1986). 'Safety attitudes of management'. *Ergonomics*, **28**, 61–67.

Harrison, E.F. (1987). *The Organisational Decision Making Process*, 3rd edn. Boston, MA: Houghton Mifflin.

Health and Safety Commission (HSC) (1992). *The Management of Health and Safety Regulations 1992*. London: HMSO.

Health and Safety Executive (HSE) (1981). *Managing Safety: A Review of the Role of Management in Occupational Health and Safety by the Accident Prevention Advisory Unit of H.M. Factory Inspectorate*. London: HMSO.

Health and Safety Executive (1991). *Successful Health and Safety Management*. HS(G) 65, London: HMSO.

Health and Safety Executive (1993). *The Costs of Accidents*. HS(G)96, London: HSE Accident Prevention Advisory Unit, HMSO.

Heller, F.A., Drenth, P.J.D., Koopman, P.L. and Rus, V. (1988). *Decisions in Organizations: A Three Country Comparative Study*. London: Sage.

HMSO (1987) *Control of Asbestos at Work Regulations*, S1 2115. London: HMSO.

HMSO (1988) *Control of Substances Hazardous to Health Regulations*, S1 1651. London: HMSO.

HMSO (1989) *The Noise at Work Regulations*, S1 1790. London: HMSO.

Institute of Chemical Engineers (IChemE) (1994). *Safety Management Systems – Sharing Experiences in Process Safety*. Rugby, Warwickshire: Institute of Chemical Engineers.

Juran, J.M. (1964). *Managerial Breakthrough*. New York: McGraw Hill.

Kjellén, U. and Baneryd, K. (1983). 'Changing local health and safety practices at work within the explosives industry'. *Ergonomics*, **26**(9), 863–877.

Koopman, P.L. and Pool, J. (1994). 'Decision making in organisations'. In *Key Reviews in Managerial Psychology, Concepts and Research in Practice* (C.L. Cooper, and I.T. Robertson, eds), Chichester: John Wiley and Sons.

Lee, T.R. (1993). 'Psychological aspects of safety in the nuclear industry'. The Second Offshore Installation Management Conference 'Managing Offshore Safety', 29 April, Aberdeen.

Lee, T.R. (1995). 'The role of attitudes in the safety culture and how to change them'. Conference on 'Understanding Risk Perception', Offshore Management Centre, Robert Gordon University, 2 February, Aberdeen.

Levitt, J.J.H. (1951). 'Some effects of certain communication patterns on group-performance'. *Journal of Abnormal and Social Psychology*, **46**, 38–50.

Noise at Work Regulations (1989). Statutory Instrument No. 1790. London: HMSO.

Pool, J. and Koopman, P.L. (1989). 'Strategic decision making in organisations: a research model and some initial findings'. The 4th European Congress on the Psychology of Work and Organisation, Cambridge, England.

Reason, J.T. (1990). *Human Error*. Cambridge: Cambridge University Press.

R.M.S. Risk Management Software (1995). Ripley, Derbyshire: Warwick I.C. Systems Ltd.

Robens, Lord (1972). *Report of the Committee 1970–1972*, Cmnd 5034. London: HMSO.

Saunders, R. and Wheeler, T. (1991). *Handbook of Safety Management*. London: Pitman.

Simon, H. (1981). *Sciences of the Artificial*. Cambridge, MA: MIT Press.

Sjöberg, L. (1995). 'Risk perception: experts and the public'. Conference on Understanding Risk Perception, 2 February, Aberdeen.

Slovic, P. (1993). 'Perceived risk, trust and democracy'. *Risk Analysis*, **13**(6), 675–682.

The Offshore Installations (Safety Case) Regulations 1992. (1992) London: HMSO.

van der Schaaf, T.W., Lucas, D.A. and Hale, A.R. (1991). *Near Miss Reporting as a Safety Tool*. London: Butterworth-Heinemann.

Walker, D. and Cox, S. (1995). 'Risk assessment: training the assessors'. *The Training Officer*, July/August, 179–181.

Walter, S. (1995). 'Health and safety on the way to being an integral part of business'. *Health, Safety and Environment Bulletin*, May 233, 6.

Waring, A.E. (1989). *Systems Methods for Managers: A Practical Guide*. Oxford: Blackwell Scientific Publications.

Whiston, J. and Eddershaw, B. (1989). 'Quality and safety – distant cousins or close relatives?' *The Chemical Engineer*, June, 97–102.

10
Safety management: strategies

Introduction

Traditional approaches to safety management (see Chapter 3) have largely concentrated their efforts on safety rules and physical environmental controls to control employee behaviour and interactions. Furthermore, this approach largely resulted from reactive approaches to accidents rather than from proactive approaches to safety. This chapter argues strongly that the effective management of the organization, through the utilization of appropriate structures, function, roles, systems and safety procedures, is of prime importance for the prevention of a poor safety performance. While Chapter 9 discussed the general responsibilities and requirements of management, this chapter focuses on organizational issues as compared to those affecting the person or their job. It also seeks to identify some of the practical concerns underpinning these approaches and provide advice (where possible) on implementation issues.

It addresses the nature of the effective organization [O] in the generic management model (see Figure 9.4). The discussion begins by exploring issues to do with the formal structure of an organization and their application to safety.

Formal organizational structures for safety management

Robbins (1983) has argued that the structure of an organization has three main components:

1 complexity;
2 formalization; and
3 centralization.

Complexity refers to the degree of differentiation and integration which exists within an organization. This view is consistent with the previous discussions of complexity theory. There are, according to Robbins (1983), at least three kinds of differentiation: vertical and horizontal differentiation, and spatial dispersion. Vertical differentiation is operationally defined as the number of hierarchical levels separating the chief executive's position from the work – systems output and is strongly related to the span of control (Mileti *et al.*, 1977). Span of control refers to the number of people reporting to any one section or manager. Horizontal differentiation refers to the degree of departmentalization and job specialization that is designed into the organization. Spatial dispersion, on the other hand, relates to a different kind of differentiation and may be defined operationally as the degree to which the organization's facilities and personnel are dispersed geographically from the main headquarters. *Formalization* refers to the extent to which jobs and tasks within an organization are standardized (Hendrick, 1987). In highly formalized organizations jobs are prescribed in detail and allow for little employee discretion. In a less formalized (employee-empowered organiza- tion) the opposite pertains. Finally, *centralization* relates to the degree that formal decision making is concentrated in an individual or high level group within the organization (see Chapter 9). All three aspects have implications for the design and operation of safety management systems.

The design and operation of safety management systems, and the safety structure within an organization, are often depicted with the aid of organization charts. A typical chart (see Figure 10.1) provides an indication of the formal relationship between the most senior management, line management, supervisors (or team leaders) and other personnel within a facility, department (etc.), or even the whole organization. Safety is a line management function (HSE, 1991), and it is important to establish how the power and authority for safety related decisions are distributed within the organization. In most organizations, the positions which are placed higher up the chart have more power and authority than those lower down (*vertical differentiation*) (in the example in Figure 10.1, this is the general manager and his or her first reports). The lines linking the positions in the chart also depict the formal channels of reporting and communication used to exercise this authority. The overall shape of the chart shows the number and breadth of levels of authority that are present in the organization. A wide flat chart is indicative of an organization with few levels of authority and a narrow tall chart reflects an organization with many.

Organization charts also show the relationship between specific jobs or roles within the organization at the same level (*horizontal differentiation*). Although some organizations include named individuals on their charts, (especially within safety policy booklets), the basic function of the chart is to represent the organizational structure in a formal sense, regardless of the particular personnel who hold the job. Drucker (1968) underlined the

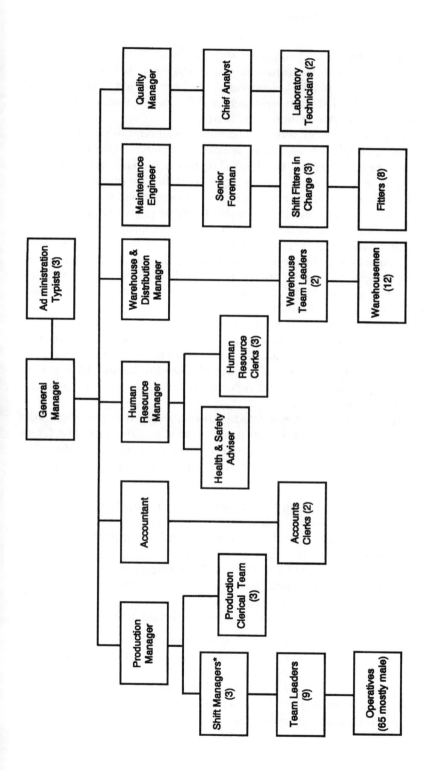

Figure 10.1 A typical organization chart

* Shift manager designated the role of incident and emergency coordinator for the site

importance of structure by arguing strongly that: 'a poor organization structure makes good performance impossible, no matter how good individual managers may be'. However, it has been argued to the contrary (Makin *et al.*, 1989) that ultimately personalities rather than roles are the important factor in organizational structures.

Organizational charts can reflect the division of labour within a particular organization through the vehicle of horizontal differentiation (see Figure 10.1). Many organizations are subdivided by function on a horizontal basis. At the same time, decision making may tend to be *centralized* and be allocated to a relatively small number of people, or alternatively it may tend to be *decentralized* (see Chapter 9). Arguably the latter is a better approach for safety management. The centralization of decision making is often a case of drawing that process both into and up the core of the organization. A reasonable balance may be struck between centralization and decentralization through application of the principle of subsidiarity which to the authors is making decisions at the lowest most appropriate level and in the most appropriate part of the organization.

A further characteristic of organizational structure is *span of control* (see above). The span of control of a manager may be made obvious on the organization chart (Arnold *et al.*, 1991) as the number of personnel who directly respond to him or her. For example, the Production Manager in Figure 10.1 would have six direct reports, including three shift managers and three clerical personnel.

There are a number of safety-related issues with regard to organizational structure; two of the most important relate to the allocation of safety-related roles, and the appropriate spans of control associated with those roles. The allocation of organizational roles for effective safety management should recognize:

1 that safety responsibility and accountability should be designated at the most senior level of the organization;
2 the formal delegation of that responsibility and accountability through line management roles and appropriate spans of control;
3 the necessary integration of line and functional management roles and the need for effective communication between role incumbents;
4 the lines of report for safety advisers; and
5 the roles and reporting lines for incident and emergency coordinators.

The second issue relates to appropriate spans of control, which in turn raises questions such as 'is the supervision appropriate and is safe working adequately monitored?' For organizations in which there is a high degree of spatial dispersion, control may be problematic and needs to be assured in some way. At a more general level, however, there are considerations of optimal structures and their impact on communication and decision making for safety.

Allocation of occupational roles for safety

Many organizations nominate a particular director or senior manager as the person who holds formal responsibility for coordinating safety- and health-related issues within the organization. Their responsibilities may include formulating organizational policy (at the primary level), ensuring that adequate control systems are in place to eliminate and control safety and health risks, monitoring performance and overseeing periodic reviews (see Table 10.1 adapted from S Cox and Fuller, 1995). They may also have a pivotal role in managing the external and political interfaces of the organization for safety (Drucker, 1994). In practice, however, responsibility for safety on a day-to-day basis is delegated through line management. Line managers, thus, have to have the necessary knowledge and competencies to fulfil such a role. The roles and competence requirements to achieve the organization's safety objectives are elaborated in Table 10.1. Three distinct roles are outlined:

1 directors and senior managers, whose purpose is to develop organizational (or corporate) policy and plans to direct and control safety risks;
2 line managers, whose purpose is to devise, execute and maintain systems to control the implementation of policy; and
3 supervisors and team leaders, whose purpose is the day-to-day implementation of safety systems to control work activities.

This emphasis on line managers' responsibilities is particularly important in securing an effective organization for safety. It had been thought that safety issues should be dealt with independently or separately from other organizational concerns to avoid them being 'sidelined' (Wallace, 1995). However, this 'independence', in the absence of a strong powerbase for the safety personnel, often meant a marginalization of safety issues until senior management were thrust (by an unprecedented accident or incident) into a response (Dawson et al., 1988; Fennell, 1988).

Functional managers, including engineers, designers and quality co-ordinators, should also have defined safety competencies and this issue is currently being addressed by many of the relevant professional bodies (see, for example, Engineering Council, 1993). The role of the safety adviser is particularly important to many organizations and is considered below.

Safety advisory roles
Safety advisers (also referred to as safety managers) occupy a key role in many organizations. They have several core responsibilities including the provision of accurate and appropriate advice to line managers and of independent advice to all employees (Saunders and Wheeler, 1991). The Lead Body on Occupational Health Safety Professionals (OHSLB) has

Table 10.1 Management roles and competence requirements for safety (adapted from S Cox and Fuller, 1995)

Occupational Roles	Directing risk management activities including health and safety	Managing health and safety activities	Supervising health and safety activities
Purpose	Develop corporate policy and plans to direct and control health and safety risks	Devise, implement and maintain systems to control the implementation of policy	Day-to-day implementation of health and safety plans and standards to control work activities
Competence requirements	1 Devise and document health and safety policy	1 Establish necessary systems to incorporate plans and performance standards	1 Implement systems and maintain development
	2 Devise corporate/ organizational plans and necessary protocols	2 Establish/operate systems for measuring health and safety performance	2 Supervise implementation of plans to achieve health and safety objectives in accordance with performance standards
	3 Measure health and safety performance	3 Establish/ operate systems for auditing and reviewing health and safety performance	3 Measure health and safety performance in accordance with measurement system
	4 Review performance and initiate beneficial change	4 Establish/ operate systems to promote a climate for the development of positive health and safety culture	4 Review health and safety performance in accordance with established system
	5 Establish the importance of a positive health and safety culture		

recently described the core competencies required of safety advisers (Royal Society for the Prevention of Accidents (RoSPA), 1995) during the launch of health and safety National and Scottish vocational qualifications (NVQs/ SVQs). These NVQs are based on standards common to general health and safety, occupational hygiene and radiological protection. The NVQs in 'occupational health and safety practice' are awarded at two levels:

1 level 3 – for practitioners in workplaces where risks are relatively straightforward; and
2 level 4 – for practitioners in workplaces where the risks are more complex.

NVQs will be awarded on the basis of assessment against standards of competence and consist of a number of 'units', each of which is further broken down into between two and four 'elements'. The element contains details of performance criteria, range statements (i.e. the full range of areas in which the unit applies) and evidence specification. Figure 10.2 (adapted from OHSLB, 1995) illustrates the detail of a single element (Element C001 – Promote a positive culture of health and safety for complex risks).

In organizational terms, the safety adviser may be perceived as a 'pivot' for safety actions and risk control. However, safety advisers do not usually hold line management positions and are thus dependent on the quality of their influencing skills, rather than formal power, to facilitate the necessary actions. Their powerbase is often derived from their apparent 'expertness' and it is therefore important that they develop and are seen to hold appropriate professional qualifications (S Cox, 1993; Dawson, 1986). One of the organizational structure requirements to support effective control of safety risks is the proximity of the safety adviser and function to senior management (Tait and Jones, 1992). Wherever the safety advisory function is located within the organization, this issue is paramount (see later). It is also important to ensure that the safety adviser keeps up to date on safety matters and current thinking (Wallace, 1995).

Small and medium organizations who do not have a full-time safety adviser may delegate this function, on a part-time basis, to a supervisor or middle manager. Alternatively, they may use external consultants in the safety adviser's role to support them in risk control.

Structure and optimal safety performance

Research into the effectiveness of various organizational structures has been mainly focused in the commercial rather than the safety domains. The classic work of Woodward (1965), which is representative of the technological determinist school of organizational theorists (Buchanan and Huczynski, 1985) is relevant here. She explored the relationship between technology and structure, and between structure and economic performance (profitability).

[1]   Performance criteria

a   Promotion of health and safety in conjunction with relevant people

b   Promotional plans and objectives respond to identified needs

c   Appropriate opportunities for contributions identified and agreed

d   Adequate details agreed with contributors and recorded

e   Comprehensive assistance provided as necessary

f   Promotions assessed and evaluated

g   Clear, concise and pertinent contributions are made to reporting and accounting for health and safety in organizational communications

[2]   Range statements

I   Relevant people: managers, supervisors, individuals with influence

II   Promotional plans and objectives are: general, related to local needs

III   Contributions: formal, informal, general, specific to a need

[3]   Evidence requirements

a   Performance

I   At least one organization-wide promotional plan plus objectives

II   Evidence of evaluation covering all types of contribution, involving all relevant people plus 3 items of communication material

b   Knowledge

Legislation, policy, promotion methods, sources of supporting materials, details of work activities including risk assessment, potential contributors' personal knowledge, organizational improvement needs, safe working practices, communication techniques, etc.

c   Sources

I   Observed performance: communications with relevant people whilst encouraging and supporting promotional activities and a review of previous supporting action

II   Products of work
- records of discussions and consultation
- general/local promotional plans
- examples of communication on health and safety matters
- examples of support and results of monitoring

Figure 10.2 Illustration of safety adviser competence requirements – Element C001 (adapted from OHSLB, 1995)

Woodward (1965) studied 100 companies in the south-east of England which used different production technologies ranging from unit and small batch production through to process and mass production systems. The organizations also ranged in size from those which employed a small number of staff to those with over 1000 employees.

Woodward (1965) identified differences in the *technical complexity* of the production technologies and processes that she was studying as she moved from prototype production to unit and small batch work to mass and finally process production systems. With this increase in technical complexity there was a corresponding increase in control and predictability in the production system. Furthermore, as the technology became more complex, the length of the chain of command increased (increased vertical differentiation), the chief executive's span of control increased, the proportion of indirect labour increased and the proportion of managers in the total workforce also increased. Overall, she found no obvious relationship between structure and economic performance, but when she examined the various categories defined by production technology a relationship became obvious. She concluded that companies which had a structure close to the norm for their category were more successful in their economic performance than those whose structure deviated from that norm.

Woodward (1965) believed that the complexity of the (production) technology determined organizational structure, and that there is 'a particular form of organization most appropriate to each technical situation' (Woodward, 1965, p.72). However, Perrow (1967) has essentially argued that the relationship between technology and structure is more complex, and is mediated by the predictability of associated work tasks and the degree of role specification that predictability allows. Buchanan and Huczynski (1985) have summarized Perrow's (1967) argument thus:

type of technology ⟶ degree of predictability of work;
degree of role specification ⟶ type of organizational structure.

Whatever the exact mechanism underpinning these relationships, the research findings illustrate the importance of technology for the design of organization structure. One of the key issues for safety, however, is the nature of adaptive, as well as non-adaptive, control systems–management as well as design. The studies cited above reflect stable organizations. There are now additional considerations which relate to de-layering (or right-sizing) within organizations and the increasing use of permanent con-tractors. It is important that an organization ensures that there is appropriate control over contractor activities and that any changes in organizational structures and roles are monitored for safety consequences. Equally, the uncertainty that can result from such changes can impact on the effective-ness of organizational structure. Several researchers (see for example, Emery and Trist, 1965; Lawrence and Lorsch, 1969; Duncan, 1972) have explored the relationship between organizational structure and uncertainty. In high uncertainty situations, it appears that it is important for the organization to have relatively low vertical differentiation, decentralized decision making and low formalization (see earlier). Such structures would appear to place, at least, aspects of control closer to those involved in the organization's day-to-day operations (cf. subsidiarity).

Tait and Jones (1992) carried out research into the location of the safety function within a representative sample of organizations in the north-west of England. They first looked at where (in the organizational structure) the health and safety function was located and then questioned safety advisers on the advantages and disadvantages of such locations. Table 10.2 illustrates

Table 10.2 Location of the safety function within a representative sample of organizations within the north-west of England (adapted from Tait and Jones, 1992)

	Autonomous	Human resource management departments	Others
All industry	49 (33%)	63 (43%)	35 (24%)
Manufacturing (large)	27 (44%)	32 (52%)	
Services (large)	10 (27%)	17 (45%)	

where the safety function was located in the organizations in the sample. Overall 33 per cent were autonomous, 43 per cent were located in human resource and other management departments, and 24 per cent in 'other' locations. Those safety advisers who were located within the human resource (or personnel) function perceived the following advantages:

1 they received good support and had access to employee records;
2 they were centrally placed for communication; and
3 they were in direct communication with the management.

However, they also perceived themselves to be remote from the shop floor, lacking in status and not having enough time for safety work. By contrast, those safety advisers who were autonomous reported an advantage in being independent but, at the same time, felt isolated from support services (Tait and Jones, 1992).

Bringing these various strands of research together, two things become obvious:

1 first, that the safety adviser or function, in order to maintain and enhance control, needs to be close to day-to-day operations (the shop floor) and thus may be particularly important in times of uncertainty and where there is high technical complexity;
2 second, that, at the same time, the safety adviser or function must cultivate strong links, informal if not formal, with senior management and appropriate strategies for influencing their thinking, philosophies and policies.

These two things must be done together, as each on its own would appear to be a recipe for failure. Their achievement might, in operational terms, be easier if the adviser or function is autonomous but, at the same time, autonomy and the associated risk of isolation might rob those involved of the necessary social support.

Safety rules – a necessary evil?

Amongst the many debates on organizations, the necessity for some degree of bureaucracy has assumed prime importance (see for example, Buchanan and Huczynski, 1985; Katz and Kahn, 1978). One of the key features of a bureaucratic organization is the pattern of rules which it maintains that are meant to guide, and hence structure, the behaviour of its members (Weber, 1947). To a great extent this rule-based approach was historically reinforced for safety by regulatory authority. Prior to the introduction of the Health and Safety at Work etc. Act 1974 (HASAWA) organizations were faced with a prescriptive regime which demanded compliance with a plethora of workplace regulations. Organizational safety rules were a logical extension of such an approach. However, the advent of goal-setting and risk management approaches (see Chapter 9) arguably challenges strict adherence to organizational safety rules but hopefully facilitates a more informed and participative approach. This logically leads to a questioning of the role of safety rules within modern organizations, and compliance with such rules.

Evidence on rule adherence and rule breaking is complex. Buckley (1968), for example, found that the number of human interactions within organizations he studied that were covered by clearly defined rules was relatively small. He also noted the existence of certain 'ground rules', but beyond that found that there was a continual process of negotiation with rules being ignored, argued and stretched, dependent on the particular situation. Human action and choice based on prior experience seemed to be important in this respect (Buckley, 1968).

THE REACTIVE
ORGANIZATION

Fear of prosecution

Standards reliant on
accident litigation

Stardards reliant on
external enforcing
agency inspections

**Health and
safety culture
based on
external
motivation**

Standards reliant
on insurance
company surveyor

Standards improved
by enforcement
notices

Corporate dictates

THE PROACTIVE
ORGANIZATION

Shares organizational
learning

Contributes to 'community'

**Health and
safety culture
based on
positive internal
motivation
proactive systems,
quality, etc.**

Communicates positively to
share–holders etc

Supports the efforts of external
enforcing agencies through
'self–regulation'

Figure 10.3 Characteristics of reactive and proactive organizations

Health and safety requires strict adherence to rules in some areas (for example, no smoking in a flammable area or evacuation of buildings in case of fire), and education and direct supervision are necessary precursors to compliance. In other areas, such a controlled regime may not be so sensible. Whatever, if a participative approach to risk assessment is adopted, it can promote the formulation of appropriate safety rules and procedures and may encourage compliance in all types of areas.

Reason and his coworkers at the University of Manchester (Reason *et al.*, 1994) have carried out field observations in a variety of industries (including oil production and exploration, rail transport and commercial aviation) and suggested that safety rules differ from other types of procedural guidance in at least two important respects:

1 the manner in which they are created (or evolve); and
2 the way that they are enforced.

Reason (Reason *et al.*, 1994; Reason, 1995) has argued that, whereas procedures governing efficiency of work tend to arise quite naturally from the nature of the prevailing technology and the tasks to which it is put (cf. Woodward, 1965; Perrow, 1967), safety rules and procedures are continually being amended to prohibit actions that have been implicated in recent accidents or incidents. Over time, these additions to the safety rules (Reason *et al.*, 1994) can become increasingly restrictive until they are finally perceived as being counterproductive. The enforcement of 'safety rules', which may have evolved in this manner, may further enhance the conflict between safety and production goals (see, for example, Reason *et al.*, 1994; S Cox, 1988; Glendon and McKenna, 1995). Flagrant and serious infringements of safe practices may well give rise to sanctions, but 'routine' violations (see Chapter 7) (Clarke, 1994; Reason, 1990) and those necessary to 'get the job done quickly' are often condoned or even requested. Clarke (1994) has suggested that 'rule-based' violations be examined in the light of the prevailing safety culture (see later).

Reason's (1995) work in this area would seem to support the earlier work of Buckley (1968) and suggest that employee involvement in the design of safety rules, combined with a proactive systems approach (rather than 'knee jerk' reactions to accidents), should support compliance.

Proactive versus reactive organizations

One of the legacies of traditional rule-based approaches to safety has been the reactive nature of a large number of organizations' management strategies (Health and Safety Commission (HSC), 1992a). The safety related motivation of a reactive organization is deemed to be largely driven by external factors (see Figure 10.3). These include:

1 fear of prosecution;
2 standards that are reliant on external agency inspections and insurance company surveyors;
3 standards that are reliant on accident litigation; and
4 corporate dictates.

By contrast, the proactive organization (see Figure 10.3) is based on a more balanced approach in which total quality is a key feature. A proactive organization for safety:

1 communicates positively and in an open and honest manner to share-holders and other stakeholders;
2 supports the efforts of external enforcing agencies through goal-setting and self-regulation; and
3 shows itself fitting to contribute to community schemes and to share organizational learning for safety.

Appropriately structured systems (see later) and effective control processes appear to be a central feature of proactive organizational strategies.

Whiston and Eddershaw (1989) have described how the principles of quality management can be applied to safety in proactive organizations (see Chapter 9). In the same vein, Reason (1995) has elaborated Hopwood's (1974) model to show how administrative controls, technical controls, social controls and self- (or person-) centred controls can be integrated to provide control of the total system. These are all necessary elements of the self-regulation depicted in Figure 10.3. Administrative and social controls are discussed in later sections of this chapter, technical controls are considered in Chapter 11 and self- or person-centred controls are addressed in Chapter 12. Supervision and the balance between imposed and self-supervision for effective social control are considered below.

Supervision

Adequate supervision is a necessary requirement to ensure effective safety. HSE (1991) have highlighted two key aspects of such supervision:

1 task management which involves the provision of direction, help, guidance, example and discipline towards meeting the required perform-ance standards; and
2 team building in which the supervisor encourages individuals to work together in pursuit of team health and safety objectives.

Levels of supervision need to be carefully considered by management. Appropriate levels will be dependent on a number of issues including legal requirements (for example, for young persons and trainees), the overall level of risk and the individual competencies of employees. However, the philosophy of empowerment is increasingly being applied to the issue of control within organizations.

Empowerment and worker involvement for safety

Current approaches to safety management in the UK (HSC, 1990; HSE, 1991) have stressed the benefits of employee involvement for safety performance. Some organizations (CBI, 1995) have taken the underlying principle to its limits and suggested that (for health and safety) the organization chart could be turned on its head and that decision making power for safety be given to those people who do the jobs. In theory this approach can be considered to be a method of ensuring that safety is the shared responsibility of all employees. This suggestion is not inconsistent with what has already been suggested in this chapter, but needs careful refining. In practice such a process needs a clear focus and requires careful facilitation (Ramsay, 1991; Saunders and Wheeler, 1991). One of the current authors (S Cox and Vassie, 1995) has designed a behavioural safety programme which is designed to facilitate employee involvement in safety related decision making. The programme is implemented by cascade through the organization. The initial stages of the programme (see Figure 10.4) are targeted at a representative group drawn from all levels of the organization including the most senior manager. After much discussion, the next stage becomes the responsibility of this team which then empowers the next generation of teams and so on throughout the organization. These subsequent teams tend to represent particular work groups or workplaces. The programme is designed to allow workers not only to input into strategies for improved hazard control within their own immediate working environment, but also to feed ideas into the total work environment through the overall programme facilitator. It has also been designed to ensure that adequate resources, including time for training and implementation, are incorporated into the programme. Full empowerment is dependent on control over adequate resources.

Formal consultation for safety

Most EU member countries have some form of worker consultation in safety and health but the UK is in the minority in giving statutory representational rights only to recognized trade unions (Gevers, 1983; Walters, 1990). The publication of the EC Framework Directive 89/391/EEC raised the possibility that the UK legislation may have to be revised to accommodate the EC demands. The UK Health and Safety Commission has not yet seen the necessity for change in general (James, 1991). However, the Offshore Installations (Safety Representatives and Safety Committees) Regulations

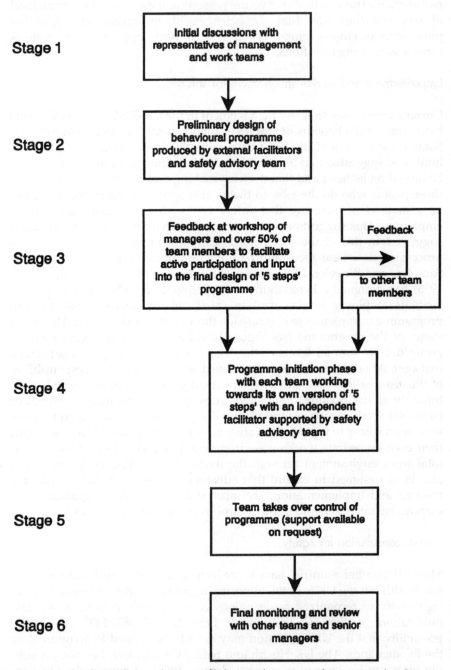

Stage 1

Initial discussions with representatives of management and work teams

Stage 2

Preliminary design of behavioural programme produced by external facilitators and safety advisory team

Stage 3

Feedback at workshop of managers and over 50% of team members to facilitate active participation and input into the final design of '5 steps' programme

Feedback

to other team members

Stage 4

Programme initiation phase with each team working towards its own version of '5 steps' with an independent facilitator supported by safety advisory team

Stage 5

Team takes over control of programme (support available on request)

Stage 6

Final monitoring and review with other teams and senior managers

Figure 10.4 Flowchart to illustrate the stages of behavioural programme for empowered work teams

1989 did adopt an elective system which entitles the workforce on an installation to nominate and elect safety representatives irrespective of trade union affiliation.

Kidger (1992) has reviewed the two contrasting approaches to employee participation in safety and, in particular, has considered the extent to which the offshore (elective) system provides a model that could be utilized throughout the whole of the UK. His arguments in favour of the 'offshore' (elective) system are that:

1 the decline in union membership since the mid-seventies has restricted employee participation rights unnecessarily; and
2 the effectiveness of employee participation should be based not only on 'rights' but also on the 'usefulness' of contribution.

Both arguments need to be considered in the light of the evidence on employee participation to determine whether representatives would be more or less effective without union support and whether this would impact on the usefulness of their contribution to the safety system. Gevers (1983) has reviewed safety representation throughout the EC and has concluded that:

> some legal backing seems to be helpful, the optimum amount of legislation being at a point where it provides for the election of worker representatives with sufficiently strong rights to allow them to function independently of management, while at the same time leaving room for a certain variety in the organizational arrangements for participation in different sectors and undertakings.

This approach suggests that organizations should make employee representation part of their culture (see later), and argues for the usefulness of the contribution made by employee representatives in an independent and participative environment. Many organizations (Walters, 1990) already operate voluntary rather than statutory safety representative schemes. In these cases, the effectiveness of the safety representative's contribution is clearly recognized and actively encouraged.

The key aspects of organizational structure and the central philosophies which are currently used in organizations have now been considered. The following sections discuss safety systems and processes in detail.

Safety management systems

It has been argued that a safety management system (SMS) is the means by which an organization can support the effective control of hazards and risks through the management process (see, for example, S Cox and Tait, 1991;

Wright, 1994). The design of an SMS is context dependent to the extent that it relates both to the nature of the prevailing technology and to the particular sector in which the organization operates. Figure 10.5 illustrates the SMS elements suggested by the Chemical Industries Association (CIA) as part of the Responsible Care Initiative (CIA, 1993). Organizations outside this sector may not require such comprehensive systems. The UK HSE (HSE, 1991) has

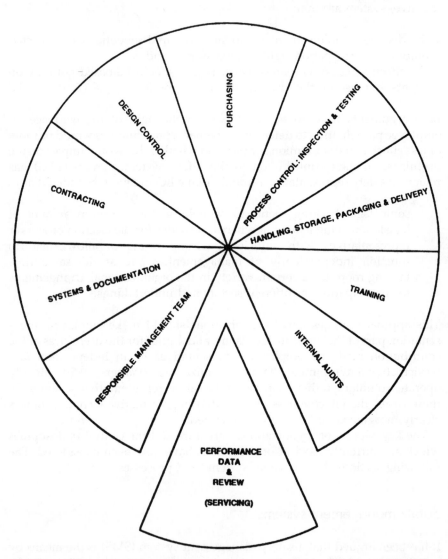

Figure 10.5 Elements of a Safety Management System as suggested by the Chemical Industries Association (CIA, 1993)

also described the main elements of successful safety management systems in terms of a cycle of continuous improvement (see Figure 9.1 in the previous chapter). This cycle can form the basis for the development of the necessary systems and procedures, and can also provide a framework for management action. The following sections discuss some of the issues surrounding the design, development and review of an SMS.

Developing an SMS

All systems comprise structural elements, processes and interconnections (see Chapter 4 for a detailed account). The safety management system is a sub-system of the total safety system and is focused on those elements, processes and interactions that facilitate risk control through the management process (Wright, 1994). The process of systems development has been described in S Cox and Tait (1991). It incorporates the complementary activities of systems analysis and planning. Various techniques have been used for such analysis and planning in the safety domain (Waring, 1989); they often utilize diagrammatic representation to support analysis. Examples include:

1 organization charts, systems maps, and influence diagrams (structures and relationships);
2 flow charts, decision sequences, and process charts (flow block and data flow); and
3 rich pictures and spray diagrams (thinking aids).

The next generation of such representations may be invested in interactive multimedia and expert systems technology.

Figure 10.6 (adapted from S Cox and Tait, 1991) describes a flowchart produced for the development of a safe system (see Checkland, 1981) for examples of the other diagrams). One of the benefits of the diagrammatic approach to the development of a safety management system is that it allows the 'developers' to engage in lateral thought and to think creatively about the system requirements (see, for example, Morgan, 1989). It also facilitates group inputs into the analytical, planning and design processes (Gevers, 1983). However, such analyses are only part of the picture, they need to be supported by sound planning to be put to good effect.

Table 10.3 outlines the key elements of an SMS (HSE, 1991) and details some strategic design and planning questions. This table is designed to support both processes. It also points up other important organizational processes for safety including selection, training and performance review. It is included as an aide memoire for managers and practitioners.

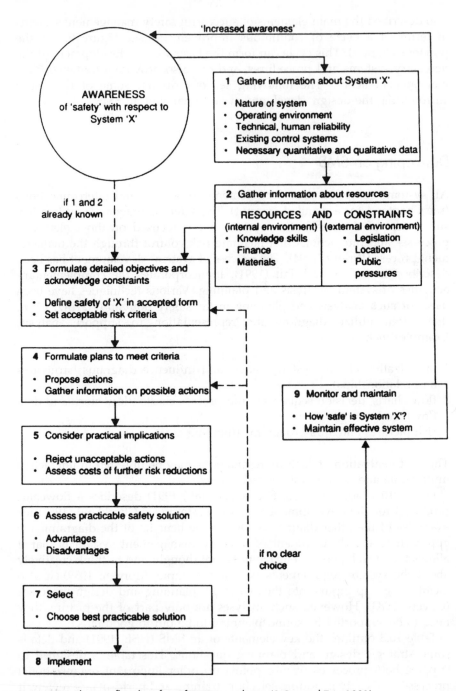

Figure 10.6 Planning flowchart for safety systems design (S Cox and Tait, 1991)

Table 10.3 Key elements of a safety management system: planning and design issues (adapted from S Cox and Fuller, 1995)

Element	Planning consideration	Implications for design
Policy	Does it convey the general intentions, approach and objectives of organization? Does it reflect critical success factors (I-J-O)? Does it incorporate a people-focused approach and has it got clarity of purpose? Is it logical and systematic and founded on sound analysis?	Is it integrated and coherent? Are senior management involved in formulation? Has it been reflected through the organization? Does it stress the importance of management control through such structures?
Organizing: Control	Can the organization improve its safety culture through securing management commitment and control? Can they plan the following tasks: • ongoing formulation of objectives; • planning, measuring and review; • sound implementation of necessary standards for jobs and individuals?	Clear allocation of management roles throughout the organization (stress importance of senior managers). Have the necessary competencies been defined (see later section)? Will personalities play a part?
Organizing: Cooperation	Has the organization formally recognized safety representatives? Are there other vehicles for cooperation – suggestion schemes, toolbox talks, etc.? Does the organization encourage near-miss reporting?	If safety representatives are in place, are they trained, motivated and able to make a positive contribution to safety? Are there design problems inherent in current systems of cooperation?
Organizing: Communication	What are the current ways in which the organization communicates safety issues (see Table 9.2)?	Can the organization design a better way? Are safety communications fully integrated within 'mainstream' communication systems (for example, team briefings, etc.)?

Table 10.3 (*Continued*)

Element	Planning consideration	Implications for design
Organizing: Competence	What are the key competencies with respect to core task skills, and safety responsibilities and skills? Is selection for safety an issue? Is there a training plan?	How are employees selected for other organizational issues – is safety a factor? Are all the key competencies defined and fully incorporated into organizational systems? Is training for safety given priority? Are there sufficient training records?
Planning and Implementing	Have risk assessment protocols been agreed? How are the output of risk assessments incorporated into safe working systems? Is there a strong and positive motivation to implement safe working practices?	Does the risk assessment system dovetail well with other organizational systems – HAZOP, HAZAN, Job Safety Analysis techniques, etc.? Are there empowered decision makers? Are line managers well trained and competent in active safety management? Is the level of supervision adequate and appropriate for safety?
Performance Measurement	Has the organization considered sufficient opportunities for proactive monitoring? Are there systems for reporting accidents, incidents and ill-health occurrences?	Are proactive monitoring opportunities considered in formulating standards and objectives? Do all personnel understand reporting criteria and are they trained to secure performance data? Are there adequate computer-based systems to record accident data?

Table 10.3 (*Continued*)

Element	Planning consideration	Implications for design
Audit and Review	Is the concept of auditing as an independent process understood? Can the organization design its own system? Are there proprietary systems available? How often should audits take place – locally, divisionally and organizationally? How should the organization handle the processes associated with safety review? Are feedback mechanisms secured?	Are there definitive training needs for auditing? Is the culture one which supports 'our' own audit process? Can proprietary audits be adapted to meet our needs? Can the organization facilitate review through improved communication and design?

Performance monitoring

Drucker's (1994) maxim, 'what gets measured, gets done', is a reminder of the importance of measuring performance if safety goals and objectives are to be taken seriously. Chapter 9 has pointed out the need to consider two broad types of measurement in the safety area:

1 proactive measurement systems which monitor the achievement of plans and the extent of compliance with standards; and
2 reactive measurement systems which monitor accidents, occupational ill-health and near miss incidents.

It has been argued (HSE, 1991) that good practice requires the development of both proactive and reactive performance monitoring. Figure 10.7 (adapted from Dupont De Nemours, 1990) illustrates how effective performance measurement can actually support improvements in safety performance. Organizations who only monitor accident statistics (Box A) are not exploiting the possibility of performance monitoring as fully as those who also monitor both strategy (Box B) and systems (Box C). Indeed, it is only by fully exploiting the measurement of safety performance that organizations can set appropriate targets.

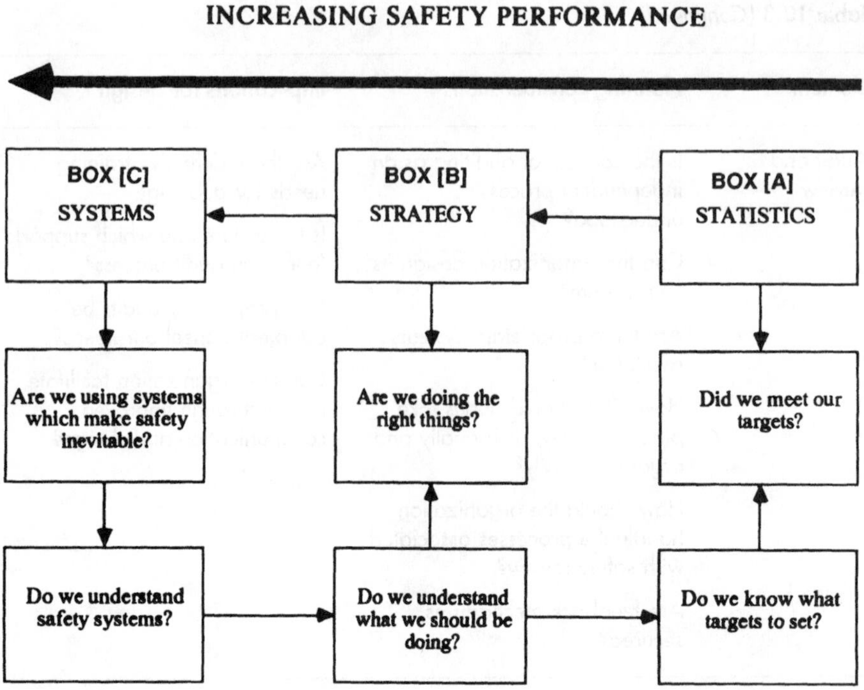

INCREASING SAFETY PERFORMANCE

Figure 10.7 The 'Dupont' model of performance monitoring

Proactive monitoring systems

A three-tier system of monitoring performance (see Table 10.4) is implicit in the current HSE guidance on safety management in higher education (HSE, 1992). The primary level is inspection, which is defined in terms of safety performance in specific work areas according to predetermined plans and against present standards in order to recognize achievement and, where necessary, to take remedial measures.

The secondary level is monitoring. Here it is defined in terms of checking the operation of the SMS, including the inspection processes. Monitoring,

Table 10.4 Inspection, monitoring and auditing systems

Workplace	Hazard control and health management processes
Level 1:	Inspection
Level 2:	Monitoring
Level 3:	Auditing

like inspection, is a process internal to the organization. The tertiary level is auditing, which is defined in terms of a (data driven) review of the efficiency, effectiveness and reliability of the total SMS including monitoring and inspection. Auditing is seen as broader based and more objective than monitoring and inspection – it may be conducted by competent agencies outside the organization.

The current authors (T Cox and S Cox, 1993) have described auditing as a process of accrediting the organization's occupational health and safety management systems and performance. Glendon and McKenna (1995) have identified at least six types of safety audits including: (a) those which cover specific topics; (b) plant and site technical audits, designed to establish that legal requirements are being met; (c) compliance audits; (d) validation audits and (e) the management safety audit (or area safety audit). Auditing should be comprehensive and designed to examine all the components over time. Safety auditing is reviewed by Kase and Wiese (1990).

The following section describes an SMS compliance audit system.

SMS review

There is a legal requirement (see Chapter 9) for some organizations to demonstrate the efficacy of their SMS (see, for example, the requirements of the Offshore Installation (Safety Case) Regulations 1992). This may be achieved through periodic audit and review. The current authors have developed methods of SMS review (see Appendix I for an illustration) which are based on data collection from three independent sources.[1] This methodology (adapted from the principles of astronomy and navigation) is known as triangulation (T Cox, 1993; Glendon and McKenna, 1995). Review data on the efficacy of SMSs may be typically acquired from the following three independent and complementary sources:

1 interviews with a representative sample of personnel throughout the organization (including both horizontal and vertical slices) and questionnaire-based surveys;
2 company documented procedures (written systems) traced through the organization to validate their application; and
3 objective data on accidents and incidents, sickness absence (etc.).

The review of the efficacy of organizational SMSs is an important aspect of performance monitoring. One of the features of such reviews is the manner

---

[1] A specimen safety management system audit protocol may be obtained from the Centre of Hazard and Risk Management (CHaRM). The reader is also directed to a review of the application of safety attitude questionnaires and structured interviews in an SMS review within a multinational oil producing company (see, for example, Alexander et al., 1994, 1995).

in which the data are presented to the 'decision makers' within the organization. In particular, the feedback of employee perceptions of safety within the organization can provide a platform for further planned improvements, and may thus impact on the prevailing health and safety culture. This theme is further developed in the next section.

Strategies for improving safety culture

Current approaches to the management of safety in organizations have tended to emphasize the importance of the prevailing culture in the overall safety performance (see earlier chapters). This section reviews strategies for possible improvements to safety culture from an organizational perspective. Further considerations of the role of individual focused processes (including attitude and behavioural change) are discussed in Chapter 12. What is considered here are the management processes that both nurture and support a positive safety culture.

Schermerhorn *et al.* (1991) have identified the following factors which are indicative of a strong organizational culture:

1 a widely shared philosophy;
2 a concern for individuals;
3 recognition of 'heroes';
4 a belief in ritual and ceremony;
5 a well understood sense of the rules (both formal and informal) and organizational expectations; and
6 a belief that what employees *do* is important to others.

Rousseau (1990) and Schein (1985) have suggested that these factors may be expressed at five levels: observable artefacts (for example company logos and annual reports), observable actions, behavioural norms, consciously expressed values and beliefs and fundamental assumptions. All of these cultural factors (and levels of expression) can be directly related to safety and map onto the characteristics of low accident plants (Lee, 1995) discussed in the previous section.

One of the strategic considerations for safety culture is that all employees share the philosophy that safety is of importance to the organization (Turner *et al.*, 1989). The existence of such a belief can be sought in the overt manifestations of safety and in any organizational 'folklore' which 'talks' about the prime movers for safety (the heroes). For example;

> our Chief Executive is so committed to safety that he has declared himself to be the safety officer for the company. Yesterday he visited our plant and the first item on the agenda was a safety tour. During

this tour he demonstrated the latest corporate approach to failsafe systems and complimented my workmates on their excellent safety record. It is obvious that safety is at the very heart of our company. (Process operative in SMS audit (anon)).

This kind of constructive observation can be fostered through management actions and an approach in which senior managers lead by example (S Cox *et al.*, 1995). Demonstration of their commitment, and that of the organization, must affect all levels of expression (Rousseau, 1990; Schein, 1985). Many organizations are taking steps to promote a positive safety culture. This has led them to consider the measurement of safety culture.

Assessment of organizational safety culture

Freytag (1990) has identified seven methods that have been used to assess organizational culture including individual interviews, group interviews, researchers' observations and interpretations of artefacts, insider's descriptions (i.e. ethnography), questionnaires, critical incident techniques and field simulations. He suggests that a variety of methods should be used at any one time to build up a picture of a particular organization. Here, one can again appeal to the principle of data triangulation.

The measurement of safety culture has been the focus of a number of reviews of safety management (see, for example, Alexander *et al.*, 1994; ACSNI, 1993; Cooper and Phillips, 1994; Donald and Canter, 1993). They each provide insight into collective values and norms for safety. Furthermore, such measurement exercises have been considered to be the first step towards cultural change (Alexander *et al.*, 1994). However, there are at least two further considerations in promoting cultural change which should be incorporated into organizational safety strategies. The first relates to the general management of change within organizations while the second is concerned with managing the organizational learning cycle.

Establishing a safety culture – incorporating change

It appears that there are three factors which should be considered when managing culture change in any area of business activity (S Cox *et al.*, 1995). These factors have been observed in the introduction of different types of safety programmes (Cohen, 1977) and include:

1 dissatisfaction with the existing state – within the health and safety arena this is usually focused on unsatisfactory accident or ill-health performance;
2 a vision of the new 'safe and healthy culture'; and
3 an understanding of how to achieve it.

First, there must be *dissatisfaction* with the existing state of affairs, either with the results or outcome measures as judged against the existing conditions of health and safety or success of the business. Dissatisfaction must have reached a critical level where employees feel (i.e. emotionally) driven or motivated to try doing something differently. It needs to be articulated in terms of expressions of discontent – 'we are no good at safety, we have too many minor accidents' or 'it's about time something was done to improve safety conditions generally'. Second, organizations need to have developed a very clear concept or vision of how they want things to be different, not simply in terms of the final result to be achieved (for example, less accidents or incidents, etc.) but, even more importantly, with an adequately detailed picture of the kinds of new arrangements (for example, operating systems, marketing, organizational structures, changed attitudes and behaviours) which will be required. If dissatisfaction with the present situation is high and the vision of the desired new state is clear, this may provide the motivation to find ways forward from one to the other. But the third condition is that no change will ever take place, even in the most desirable of circumstances, unless the person involved can clearly see some practical actions which they can accomplish to make it happen.

These three preconditions for change, assuming that each one is present to a significant degree, together form a total force for change. However, the change will only take place if this total positive force is sufficiently great to outweigh all the negative restraining forces, in other words the costs of making the change. These costs might be financial or physical, but the main resistance to change is nearly always the cost in human terms either to oneself or other people (Szilagyi and Wallace, 1990). Most people tend to be apprehensive about change affecting their lives, particularly when this is driven by other people and when they themselves feel they have little or no control or influence over it. Szilagyi and Wallace (1990) have identified the following reasons for resistance to change:

1 fear of economic loss;
2 potential social disruption (most employees value the social relationships developed at work);
3 inconvenience (changes in working practices for safety would invariably involve some degree of change);
4 fear of uncertainty; and, finally,
5 resistance from groups (if change is imposed that threatens existing group norms or sense of importance, it is likely to meet with group resistance).

Managers and change agents need to consider all of these issues in planning culture change. There are several questions which managers need to answer when changing safety culture; these are presented in Table 10.5.

Table 10.5 Changing health and safety culture

---

1 a What is the person's/group's degree of dissatisfaction with the present situation?
   b What actions can I ethically take to endeavour to raise this to a critical level?

2 a How clear is the concept of their desired future state?
   b How can I help to sharpen this?

3 a What practical steps are available/feasible for them to work towards their goal?
   b How can I usefully share my thoughts on this with them?

4 a What are likely to be the costs in human terms of this change?
   b How might these costs be reduced?

Organizational development

Organizations should, if the conditions are right, naturally evolve and again, if the conditions are right, develop in complexity while maintaining some degree of stability. Unfortunately, the conditions are not always right. There is, therefore, a need to think carefully about enhancing the process of evolution by deliberate intervention. This is essentially what organization development is about. While the change strategies discussed above have focused on safety issues, organizational development, as discussed here, is more general and concerned largely with improving the health of the organization.

In Chapter 5 it was argued that employees of healthy organizations are less likely to be dissatisfied, disaffected and distracted than those of unhealthy organizations; they are less likely to experience work-related stress (see chapter 8). Their behaviour is, for this reason, less likely to be unsafe. In addition, a good enough problem solving environment will allow answers to questions of safety to be arrived at by the work group, and a good enough development environment will ensure that the members of that group are adequately informed and competent in relation to both their work tasks and safety. Finally, a good enough task environment will allow employees to complete their tasks effectively and safely, and also meet any other safety challenges well enough. Various different techniques have been employed to promote organizational development; many use the principles outlined above for the planning and management of change. Neither change strategies nor organizational development will work in the longer term if the organization does not have the capacity to learn.

Organizational learning

One of the characteristics of low accident plants (discussed in Chapter 9) was their focus on organizational learning (Lee, 1995). For safety, this type of learning means that an organization deliberately collates, analyses and disseminates all its performance data, including its accident and incident data, so that the whole organization and its employees may learn from the incidents that have occurred. Organizational learning has been formally defined (Dixon, 1994) as 'the intentional use of learning processes at the individual, group and system level to continuously transform the organization in a direction that is increasingly satisfying to its stakeholders'. She has further postulated that effective organizational learning is an 'active' rather than a 'passive' process. This process (Dixon, 1994) may be described by an organizational learning cycle (see Figure 10.8). This cycle incorporates the following four steps:

1 widespread generation of information (this assumes that comprehensive data is available for dissemination);
2 integration of new/local information into the organizational context;
3 collective interpretation of information; and
4 having authority to take responsible action based on the interpreted message.

Most organizations gather extensive amounts of information in relation to the safety and reliability of systems. These data include: accident reports to employees and non-employees, lost time accident statistics, significant event records and records of near misses and dangerous occurrences. Certain industries will also collect safety-related data which is unique to their context, for example process industries collate reaction data, rail networks collect data on signals passed at danger, etc. Each of these organizations will expend considerable efforts in managing the data collection process. But is this effort translated into effective learning?

Lucas (1989) has highlighted five common problems with safety related databases which are:

1 technical myopia (most approaches are orientated towards 'hardware' failures);
2 an action orientated approach which considers 'what' happened rather than 'why';
3 a focus on events rather than causation;
4 dependent on consequences – more serious problems are recorded – near misses are not taken seriously; and
5 variable quality.

Toft and Reynolds (1994) have strongly argued that organizations are not effectively using system-failure data to prevent disasters through appropriate feedback systems. For example, after the sinking of the *Herald of Free Enterprise*, previous incidents came to light where other RO-RO ferries had

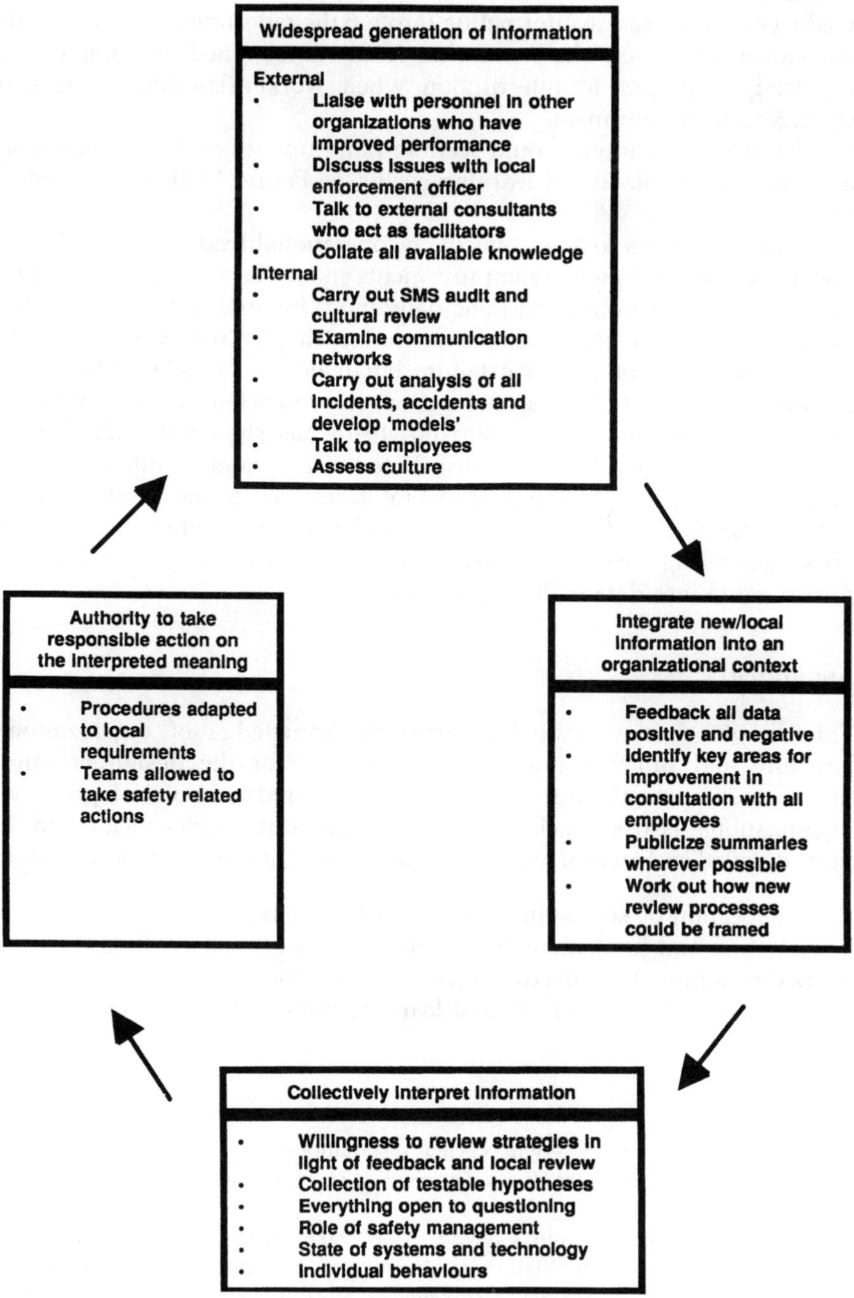

Widespread generation of information

External
- Liaise with personnel in other organizations who have improved performance
- Discuss issues with local enforcement officer
- Talk to external consultants who act as facilitators
- Collate all available knowledge

Internal
- Carry out SMS audit and cultural review
- Examine communication networks
- Carry out analysis of all incidents, accidents and develop 'models'
- Talk to employees
- Assess culture

Authority to take responsible action on the interpreted meaning
- Procedures adapted to local requirements
- Teams allowed to take safety related actions

Integrate new/local information into an organizational context
- Feedback all data positive and negative
- Identify key areas for improvement in consultation with all employees
- Publicize summaries wherever possible
- Work out how new review processes could be framed

Collectively interpret information
- Willingness to review strategies in light of feedback and local review
- Collection of testable hypotheses
- Everything open to questioning
- Role of safety management
- State of systems and technology
- Individual behaviours

Figure 10.8 An organizational learning cycle for safety

left Zeebrugge with their bow doors open. Clarke (1993) describes a feedback loop for safety information through the reporting of hazards by the workforce. However, she notes that feedback obtained through hazard reporting is subject to interruption when workers perceive a lack of management commitment.

All these observations would seem to point up failures in two necessary areas of the organizational learning cycle (see Figure 10.8) with respect to safety.

For organizations to learn effectively, operational feedback must first be positively managed and any improvements should be actively incorporated into the system. Second, and potentially more important, the organization needs to accept that 'errors' in the system (see Chapter 3) may occur not only as a consequence of technical failure but may also be related to human failures. Furthermore, human failures are not restricted to the 'sharp end'. Evidence on human failures (Reason, 1990) has shown that the human causes of major accidents are distributed very widely, both within the system as a whole and often over several years prior to the event. Evidence for 'organizational errors' is widely available and techniques for mapping error generating mechanisms (see Chapter 3, Groeneweg, 1994) can be incorporated into data gathering systems.

Summary

This chapter has attempted to translate knowledge of organizational structure and function into recommendations for the design of safety management systems. The previous chapter explored some of the higher level organizational issues which impinge on safety; this chapter focuses on the more detailed and tactical level. It provides a weighty set of suggestions for:

1 the allocation of key occupational roles for safety;
2 the nature and implementation of safety management systems;
3 the development of effective safety cultures; and
4 organizational development and learning in the safety domain.

References

Advisory Committee for Safety in Nuclear Installations (ACSNI) (1993). *Third report: 'Organising for Safety' of the Human Factors Study Group of ACSNI*, Sudbury, Suffolk: HSE Books.

Alexander, M., Cox, S.J. and Cheyne, A. (1994). 'Safety culture within a UK organisation engaged in Offshore Hydrocarbon Processing'. In *Proceedings of the IVth Conference on Safety and Well-Being at Work, November 1–2, Loughborough University of Technology* (A. Cheyne, S.J. Cox, and K. Irving, eds) pp. 67–74, Loughborough: Loughborough University.

Alexander, M., Cox, S.J. and Cheyne, A. (1995). 'The concept of culture within a UK offshore organisation'. In *Understanding Risk Perception Conference, The Robert Gordon University, 2 February, Aberdeen.*

Arnold, J., Robertson, I.T., and Cooper, C.L. (1991). *Understanding Human Behaviour in The Workplace.* London: Longman.

Buchanan, D.A. and Huczynski, A.A. (1985). *Organizational Behaviour. An Introductory Text.* Englewood Cliffs, NJ: Prentice Hall International.

Buckley, W. (1968). 'Society as a complex adaptive system'. In *Modern Systems Research for the Behavioural Scientist* (W. Buckley, ed) pp. 490–513, Aldine Publishing Company.

Confederation of British Industry (CBI) (1995). *The Force Is With You – Harnessing People Power For Total Safety.* Scunthorpe: Outtakes Video Communications.

Checkland, P.B. (1981). *Systems Thinking, Systems Practice.* Chichester: John Wiley.

Chemical Industries Association (CIA) (1993). *Guidance on Safety, Occupational Health and Environmental Protection.* February, London: CIA.

Clarke, S. (1993). 'Organisational communication and its effects on train drivers. attitudes towards safety' (Ph.D. thesis) Manchester: University of Manchester.

Clarke, S. (1994). 'Violations at work: implications for risk management'. In *Proceedings of the IVth Conference on Safety and Well-Being at Work, November 1–2, Loughborough University of Technology* (A. Cheyne, S Cox and K. Irving, eds) pp. 116–124.

Cohen, A. (1977). 'Factors in successful safety programmes'. *Journal of Safety Research,* 9, 168–178.

Cooper, M.D. and Phillips, R.A. (1994). 'Validation of a safety climate measure'. *Occupational Psychology Conference of the British Psychology Society, 3–5 January, Birmingham.*

Cox, S. (1988). 'Attitudes to safety in a multinational gas company'. Unpublished M.Phil. Thesis, University of Nottingham, Nottingham.

Cox, S. (1993). 'Safety education in universities in Europe'. In *Proceedings of the Second IUPAC Workshop on Safety in Chemical Production* (The Chemical Society of Japan, ed). Yokohama, Japan, 31 May–4 June, 48–54.

Cox, S. and Vassie, L.H. (1995). *Behavioural Safety Toolkit.* Loughborough: Centre for Hazard and Risk Management, University of Loughborough.

Cox, S. and Fuller, C.W. (1995). *Confidential Client Report on Health and Safety Systems Audit.* Loughborough: Centre for Hazard and Risk Management.

Cox, S.J. and Tait, N.R.S. (1991). *Reliability, Safety and Risk Management.* Oxford: Butterworth Heinemann.

Cox, S., Janes, W., Walker, D. and Wenham, D. (1995). *Office Health and Safety Handbook.* Croydon, Surrey: Tolley.

Cox, T. (1993). *Stress Research and Stress Management: Putting Theory to Work.* Sudbury, Suffolk: HSE Books.

Cox, T. and Cox, S. (1993). *Psychosocial and Organizational Hazards: Monitoring and Control.* European Series in Occupational Health no. 5. Copenhagen, Denmark: World Health Organization (Regional Office for Europe).

Dawson, S. (1986). 'The training and professional development of workplace health and safety specialists'. In *Education and Training in Health and Safety: Current Problems and Future Priorities – Proceedings of a Conference held by the British Health and Safety Society,* London, June. 7–20.

Dawson, S., Willman, P., Clinton, A. and Bamford, M. (1988). *Safety at Work; The Limits of Self Regulation.* Cambridge: Cambridge University Press.

Dixon, N. (1994). *The Organizational Learning Cycle: How We Can Learn Collectively.* London: McGraw-Hill.

Donald, I. and Canter, D. (1993). 'Psychological factors and the accident plateau'. *Health and Safety Information Bulletin,* **215**, November, 5–12.

Drucker, P.F. (1968). *The Practice of Management.* London: Pan Books Ltd. in association with William-Heinemann Ltd.

Drucker, P.F. (1994). *Management: Tasks, Responsibilities, Practices.* Oxford: Butterworth-Heinemann.

Duncan, R.B. (1972). 'Characteristics of organizational environments and perceived environmental uncertainty'. *Administrative Science Quarterly,* **17**, 313–27.

Duncan, W.J. (1981). *Organizational Behaviour,* 2nd edn Boston, MA: Houghton Mifflin.

Dupont De Nemours (1990). *Safety at Dupont.* Dupont De Nemours (Deutschland) Gmbh, Safety Management Services, Europe Postfach 1393, 4700 Hanm 1.

Emery, F.E. and Trist, E.L. (1965). 'The causal texture of organizational environments'. *Human Relations,* **18**, 21–3.

Engineering Council (1993). *Engineers and Risk Issues: Code of Professional Practice.* London: Engineering Council.

Fennell D. (1988) *Investigation into the King's Cross Underground Fire.* Department of Transport. London: HMSO.

Freytag, W.R. (1990). 'Organizational culture in psychology'. In *Organizations: Integrating Science and Practice* (K.R. Murphy and F.E. Saal, eds), Hillsdale, NJ: Lawrence Erlbaum Associates.

Gevers, J.K.M. (1983). 'Worker participation in health and safety in the EEC'. *International Labour Review,* **122**(4), 411–28.

Glendon, A.I. and McKenna, E.F. (1995). *Human Safety and Risk Management.* London: Chapman and Hall.

Groeneweg, J. (1994). *Controlling the Controllable.* The Netherlands: DSWO Press, Leiden University.

Health and Safety at Work etc. Act (1974). London: HMSO.

Health and Safety Commission (HSC) (1990). *Advice to Employers. Leaflet HSC3.* London: HMSO.

Health and Safety Commission (1992a). *The Management of Health and Safety at Work: Guidance on Regulations.* London: HMSO.

Health and Safety Commission (1992b). *The Offshore Installations (Safety Case) Regulations.* London: HMSO.

Health and Safety Executive (HSE) (1991). *Successful Health and Safety Management, HS(G)65.* London: HMSO.

Health and Safety Executive (1992). *Health and Safety Management in Higher and Further Education: Guidance on Inspection, Monitoring and Auditing.* London: HMSO.

Hendrick, H.W. (1987). 'Organizational design'. In *Handbook of Human Factors* (G. Salvendy, ed) pp. 470–494. New York: John Wiley and Sons.

Hopwood, A.G. (1974). *Accounting Systems and Managerial Behaviour.* Hampshire: Saxon House.

Investigation into the Kings Cross underground fire (Fennel Report, 1988). London:

HMSO.

James, P. (1991). 'HSC proposals on the Framework and Temporary Workers Directives'. *Health and Safety Information Bulletin*, **192**, 2–7.

Kase, D.W. and Wiese, K.J. (1990). *Safety Auditing: A Management Tool*. Van Nostrand Reinhold.

Katz, D. and Kahn, R.L. (1978). *The Social Psychology of Organisations*. New York: Wiley.

Kidger, P. (1992). 'Employee participation in occupational health and safety: should union-appointed or elected representatives be the model for the UK?' *Human Resource Management Journal*, **2**(4), 21–35.

Lawrence, P.R. and Lorsch, J.W. (1969). *Organization and Environment*. Hamewood, IL: Irwin.

Lee, T.R. (1995). 'The role of attitudes in the safety culture and how to change them'. In *Understanding Risk Perception Conference*, The Robert Gordon University, 2 February, Aberdeen.

Lucas, D.A. (1989). 'Collecting data on human performance: going beyond the "what" to get at the "why"'. In *Proceedings of a Workshop on Human Factors Engineering: A Task Orientated Approach*, Noordwijk, Netherlands: ESTEC.

Makin, P.J., Cooper, C.L. and Cox, C. (1989). *Managing People at Work*. London: Routledge in association with the British Psychological Society.

Mileti, D.S., Gillespie, D.F. and Haas, J.E. (1977). 'Size and structure in complex-organizations'. *Social Forces*, **56**(1), 208–217.

Morgan, G. (1989). *Creative Organisation Theory, A Resource Book*. Finsbury Park, London: Sage Publications.

OHSLB (1995). *Occupational Health and Safety Practice Qualification (Health and Safety Specific With Levels 3 and 4)*. Sudbury, Suffolk: HSE Books.

Perrow, C. (1967). *Organizational Analysis: A Sociological View*. London: Tavistock.

Ramsay, H. (1991). 'Reinventing the wheel? A review of the development and performance of employee involvement'. *Human Resource Management Journal*, **30**.

Reason, J. (1990). *Human Error*. New York: Cambridge University Press.

Reason, J. (1995). 'A systems approach to organizational error'. *Ergonomics*, **38**(8), 1708–1721.

Reason, J., Porter, D. and Free, R. (1994). *Bending the Rules: The Varieties, Origins and Management of Safety Violations*. Leiden: Rijks Universiteit.

Robbins, S.R. (1983). *Organization Theory: The Structure and Design of Organizations*. Englewood Cliffs, NJ: Prentice-Hall.

Rousseau, D.M. (1990). 'Assessing organizational culture: the case for multiple methods'. In *Organizational Climate and Culture*. San Francisco: Jossey-Bass.

The Royal Society for the Prevention of Accidents (RoSPA) (1995). RoSPA Health and Safety Congress, June, National Exhibition Centre, Birmingham.

Saunders, R. and Wheeler, T. (1991). *Handbook of Safety Management*. London: Pitman.

Schein, E.H. (1985). *Organizational Culture and Leadership*. San Francisco: Jossey-Bass.

Schermerhorn, J.R., Hunt, J. and Osborn, R. (1991). *Managing Organizational Behaviour*. New York: J. Wiley and Sons.

Szilagyi, A.D. and Wallace, M.J. Jnr (1990). *Organizational Behaviour and Performance*, 5th Edn. Glenview, IL: Scott, Foresman.

Tait, N.R.S. and Jones, S. (1992). 'Practice makes...?' *Journal of Occupational Health and Safety*, January, 35–39.

Toft, B. and Reynolds, S. (1994). *Learning from Disasters: A Management Approach*. Oxford: Butterworth-Heinemann Ltd.

Turner, B.A., Pidgeon, N.F., Blockley, D.I. and Toft, B. (1989). 'Safety culture: its position in future risk management'. Paper to *Second World Bank Workshop on Safety Control and Risk Management*, Karlstad, Sweden.

Wallace, I. (1995). *Developing Effective Safety Systems*. Rugby: Institution of Chemical Engineers.

Walters, D.R. (1990). *Worker Participation in Health and Safety – A European Comparison*. London: Institute of Employment Rights.

Waring, A.E. (1989). *Systems Methods for Managers: A Practical Guide*. Oxford: Blackwell.

Weber, M. (1947). *The Theory of Social and Economic*, translated by A.M. Henderson and T. Parsons (eds). New York: Oxford University Press.

Whiston, J. and Eddershaw, B. (1989). 'Quality and safety – distant cousins or close relatives?' *The Chemical Engineer*, June, 97–102.

Woodward, J. (1965). *Industrial Organization: Theory and Practice*. London: Oxford University Press.

Wright, M.S. (1994). 'A review of safety management system approaches to risk reduction'. Paper presented at *Risk Assessment and Risk Reduction Conference*, 22 March, Aston University, Birmingham.

11
Managing the work environment: the design of safe work

Introduction

The previous chapter has considered organizational strategies and issues that are important for the management of safety. It discussed those relating to organizational structure, roles and responsibilities, safety management systems and the prevailing safety culture. These factors provide the context for the design of safe systems of work; together they describe the social and task environments in which jobs are completed. This chapter is concerned with the design of safe work. It builds on the earlier discussions of jobs and tasks (see Chapter 6). The main aims of this chapter are to:

1 discuss the principles underpinning the design of safe work in the light of current psychological and ergonomic knowledge;
2 explore the application of job and task analyses (including job safety analysis) to risk assessment in the design of safe work;
3 outline the importance of the physical environment and associated tools and equipment with particular references to the principles of reliability in the design of safety critical systems;
4 consider the issues surrounding the organization of working hours and those relating to shiftwork; and
5 discuss practical approaches to the prevention of human error which focus on job design.

Taken together, these sections should provide an understanding of the design and subsequent implementation of safe systems of work. However, the reader is reminded that such systems need to take account of all the safety factors that have been raised in terms of the model of the 'person in their job in their organization'.

Safe systems of work

The concept of a safe system of work is fundamental to current thinking about safety management. It has been variously defined within both the legal and technological literature on safety. The authors' preferred definition derives from the UK Health and Safety Executive's (HSE) publication *Dangerous Maintenance* (1987); a study of maintenance accidents in the chemical industry. It defines a safe system of work as:

> a method of doing a job which eliminates identified hazards, controls others and plans to achieve the controlled completion of the work with the minimum risk.

This definition is consistent with the hierarchy of safety protective measures which is outlined in both UK and European law (see, for example, the

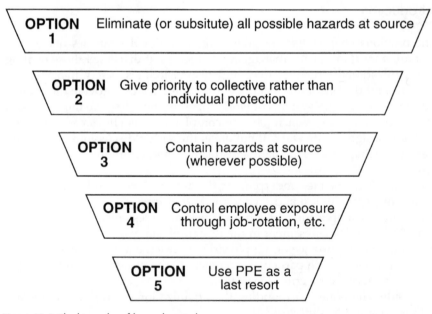

Figure 11.1 The hierarchy of hazard control

Framework Directive 89/391/EEC). This hierarchical approach demands that controls should be prioritized in accordance with the following principles (see Figure 11.1):

1 wherever possible hazards should either be eliminated or substituted at source;

2 collective (rather than individual) measures of protection should be given priority;
3 hazards should be contained at source by physical (usually engineering) methods wherever possible;
4 where physical control measures are not a feasible option, employee exposures should be reduced through other methods (for example, through job rotation); and finally
5 short term methods such as the utilization of personal protective equipment (PPE) should only be used as a last resort. If PPE is used it should comply with the appropriate regulations.

These principles need to be taken into account when control options are selected in the risk assessment process (S Cox, 1992). The components of a safe system of work which would be assured in the assessment process are described in Table 11.1. Many of these components are considered in later sections of this chapter. The first section of this chapter, however, further explores the job design process in the light of the psychology of safety outlined in Part 2 and of current approaches to workplace ergonomics and design.

Table 11.1 Components of a safe system of work

1	Well designed tasks
2	Safe plant machinery and equipment
3	Materials and substances which are adequately assessed and controlled
4	Immediate safe working environment(s)
5	People (well trained and effectively supervised)
6	Positive safety culture

Safety by design: process issues

The design of jobs using principles derived from engineering, psychology and ergonomics is a key factor in the management of safety. Experts in ergonomic job design (Rodgers et al., 1986) have observed that despite the advantages offered by a 'safety by design' approach, the application of such an approach is still not widespread throughout organizations. They suggest (Rodgers et al., 1986) that this apparent apathy and reluctance may have arisen as a result of the following reasons:

1 a failure to recognize how poorly designed jobs can impact on both safety and productivity;
2 a lack of practical guidelines that can be applied to the design of safe jobs;
3 a paucity of practical information and the unavailability of both formal documented studies and informal anecdotes; and finally
4 an acceptance of so-called 'hard jobs' and an availability of people to do them. This approach has been particularly prevalent in the heavy industries and is also associated with jobs that are found in the construction sector.

There are at least three primary arguments why organizations should incorporate ergonomic and psychological principles into job design (Rodgers *et al.*, 1986). These include enhanced productivity: Table 11.2 (adapted from Rodgers *et al.*, 1986) illustrates two job design factors and the nature of their effects on productivity and safety. Other immediate benefits of effective job design include the reduction in the number of problematic tasks and the possibility of accommodating a broader spectrum of workers in the available jobs. In the longer term there is also the potential to reduce harm from chronic exposure to ergonomic and psychosocial hazards (T Cox and S Cox, 1993). For example, lack of control by the worker (often a consequence of high paced work, see Table 11.2) has been shown to reduce job satisfaction and to affect the quality of the work output (T Cox, 1985; Sen *et al.*, 1981).

Table 11.2 Job design factors that contribute to loss in productivity (adapted from Eastman Kodak Company, 1986)

Job design factor	Workplace example	Mechanism of lost productivity and potential 'hazards'	Design improvement
High-paced work	Inspection of fast moving frozen food on conveyors whose speed cannot be controlled by operator	Poor quality product as rejects are not detected. Potential stress-related problems through lack of control	Provide some control for the 'inspector' by unlinking the workstation from the conveyor
Static muscle loading	Continuous holding of an object or work tool. Possibly stooping or crouching repeatedly	Local muscle fatigue which limits time on the task to relatively short time	Provide holding fixtures. Design workstation to avoid low clearances that require awkward postures

Other human factors specialists have recognized the importance of practical information for job design. For example, Graves and Pethick (1987) have provided a model for the design process which identifies information sources from both the immediate working environment and the wider organization. Figure 11.2 (adapted from Graves and Pethick, 1987) illustrates the sources of formal and informal information available within the organization and that within the wider environment which may be

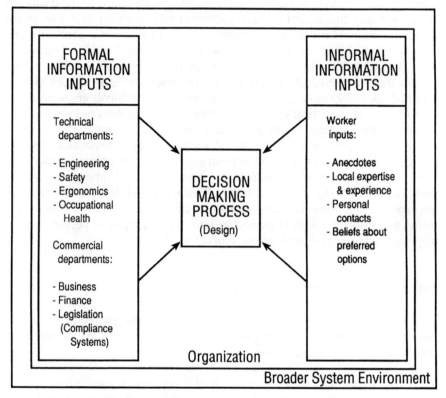

Figure 11.2 Formal and informal information inputs into the design process

incorporated into the design process. The formal inputs are from technical departments and include the appropriate engineering, ergonomic and safety knowledge. The informal sources include worker inputs and local expertise. The latter often reflect the prevailing attitudes and work experiences of the job holders and are further discussed in the next chapter. The development of information networks and databases to suit particular business environments (for example, MHIDAS and HAZDATA) should facilitate the design process. However, such networks and databases can only support safe designs if:

1 organizations have the necessary expertise and resources to access them; and, equally important,
2 they are prepared to collect and provide incident data to build up the database.

The design process involves decision making at a number of levels. These range from broad discussions on the design of the overall system to the detailed decisions and specifications for a workstation layout. The considerations in Chapter 9 as to the locus and style of decision making and the way it is integrated into line activities have particular relevance for decisions in the safety domain. Many large organizations produce generic protocols and standards which both incorporate the necessary technical and legal standards and allow for 'local' modifications. Such standards are usually held centrally within organizations and are the basis of written safe systems of work. Graves and Pethick (1987) have argued that whereas most protocols are formulated on the basis of industrial engineering data, such data does not always take account of the workers' capabilities in a consistent and systematic manner. As a consequence, the way in which the individual is incorporated into the design of the work system evolves from experience and opinion rather than from the standpoint of scientific data on worker capabilities (Graves and Pethick, 1987; Rodgers *et al.*, 1986). Person-centred (or person-focused) approaches to job and task design redress this balance and address people issues as a priority. The next section outlines the principles of person-centred approaches to the design of safe systems of work.

Person-centred approaches to job design

McCormick (1983) has discussed the process of person-centred job design in his text on *Human Factors in Engineering and Design* in three stages (its focus, objectives and central approach):

1 the focus relates to the consideration of 'people' in the design of objects, facilities and environments that people 'use' in all aspects of their lives (including their working life);
2 the objectives of person-centred approaches in this design process are twofold and include both the enhancement of functional effectiveness and welfare and safety; and
3 the central approach is the systemic application of relevant information about human characteristics and behaviour to the design of objects, facilities and environments that people 'use'.

Meister (1989) has described the application of these principles in his definition of human factors as the 'the appreciation of how humans accomplish work related tasks in the context of human-machine system operation and how both behavioural and non-behavioural variables affect that accomplishment'. Both McCormick's (1983) and Meister's (1989) approaches illustrate the breadth of this area and place fundamental importance on interactions between factors. They have also emphasized the application of appropriate 'human performance' data to the design process. These data are derived from a number of sources (see, for example, Boff and Lincoln, 1988) and involve a variety of disciplines including ergonomics, physiology, anthropometry, psychology and engineering design. Person-centred design is thus a multidisciplinary activity which involves fitting the job or task to the individual so as to maximize task performances and to

Table 11.3 Major considerations in the design of safe jobs (HSE, 1989)

---

1 Identification and comprehensive analysis of critical tasks and appraisal of likely errors

2 Evaluation of required 'operator' decision making and balance between person and machine inputs and requirements

3 Application of ergonomic principles to the design of person–machine interfaces including displays of plant and process information, control devices and panel layouts

4 Design and presentation of procedures and operating instructions

5 Organisation and control of the working environment, including the workspace, prevailing environment (lighting, noise, humidity, etc.) and access for maintenance

6 Provision of correct tools and equipment, including any necessary safety equipment

7 Scheduling of work patterns, including shift organisation, control of fatigue and stress

8 Efficient communications at all times, particularly in emergencies

---

ensure a safe outcome. The HSE human factors in industrial safety publication (HSE, 1989) stresses the importance of person-centred approaches to job design. Table 11.3 elaborates the eight major considerations highlighted in the HSE (1989) publication in relation to jobs. Several of these are discussed in later sections of this chapter.

'Person–machine' interface and systems

Part of the process of fitting the job or task to the person may be a consideration of the 'person–machine' interface (PMI). By linking the 'machine' to the person in this way (and in this direction (see Figure 11.3)) a relationship is established between the two system components. The machine displays information to the person (or operator) via the operator's sensory apparatus (see Chapter 7 and Figure 11.3). The operator uses the appropriate controls to affect the machine. Oborne (1995) has illustrated the concepts underpinning the 'person–machine' interface or 'person–machine' loop through the process of driving a car:

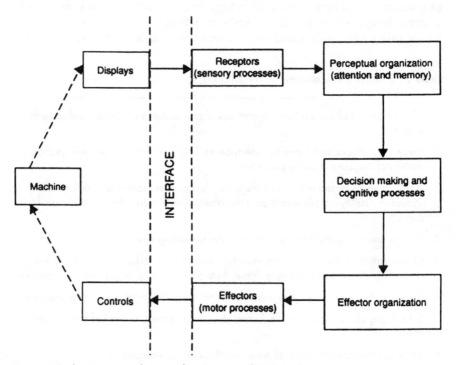

Figure 11.3 The person–machine interface (S Cox and Tait, 1991)

In order to drive a car safely and efficiently along a road, a relationship needs to be established between the driver (the person) and the vehicle (the machine) such that any deviation from the prescribed path is displayed back to the driver via the visual (and sometimes auditory) sensory mechanisms (see Chapter 7). In this way information will pass from the machine to the person and back to the machine in a closed information-control loop. Effective design incorporates the strengths and weaknesses of the human information processing capabilities discussed in earlier sections.

Although examples of single 'person–machine' loops can be seen in a range of different work situations, it is more usual for combinations to occur which increase the complexity of interactions. The challenge for system designers today is to both preserve and enhance the effective operation of these types of complex loops. Traditional PMI approaches also face a further challenge. Oborne (1995) has argued that although individual operators and their working systems should operate in close harmony, in practice operators and 'working systems' are not equal partners and people (rather than machines) should be at the centre of the design process. Furthermore, Oborne (1995) has identified the following essential features for the design of safe and effective person–machine systems:

1 *Purposivity* – the technology needs to reflect the actual use to which it is put (not the perceived use).
2 *Anticipation and prediction* – these follow on from the concept of purpose; they concern how the system is operated and controlled. For example, the way in which information is displayed to an operator should be such that they can 'see' the results of their actions before they are carried out.
3 *Interest and boredom* – this feature relates to the stimulation and interest of the operator and stems from the source of the activity. Increased interest leads to a lowered likelihood of boredom and subsequent reduction in errors.
4 *Control and autonomy* – the importance of these concepts is well recognized in organizational psychology – control (real or perceived) over the situation is paramount and reduces the uncertainty of the outcome.
5 *Responsibility and trust* – a central aspect of the person-centred approach is that individuals act with responsibility when interacting with the system. Since this responsibility is towards the successful outcome of the goals the information must be of the kind and nature necessary to facilitate the desired outcome. Any information received must also be trusted by the operator.

Allocating functions between person and machine

One of the most important problems in person–machine systems design is the allocation of functions between the operator and the machine. Which tasks should be assigned to people and which to machines? Or what type of tasks should humans be operating in a person–machine system? On what basis should this task allocation be decided? The most obvious way to separate functions, given that constraints such as cost–benefit have been considered, is on the relative capabilities and limitations of the two components. Such an approach was attempted by Fitts (1951) in drawing up his now famous list itemizing the relative advantages of persons and machines. An updated version of the so-called Fitts List as modified by Singleton (1974) is shown in Table 11.4. It identifies a number of properties against the relative performance characteristics of the human

Table 11.4 An updated version of the Fitts List (from Singleton, 1974)

Property	Machine performance	Human performance
Speed	Much superior Consistent at any level Large constant standard forces and power available	Lag one second 2 horsepower for about 10 secs 0.5 horsepower for a few mins 0.2 horsepower for continuous work over a day
Consistency	Ideal for routine, repetition and precision	Not reliable – should be monitored Subject to learning and fatigue
Complex activities	Multi-channel	Single channel Low information throughout
Memory	Best for literal reproduction and short term storage	Large store multiple access Better for principles and strategies
Reasoning	Good deductive Tedious to reprogramme	Good inductive Easy to reprogramme
Computation	Fast, accurate Poor at error correction	Slow Subject to error Good at error correction
Input	Can detect features outside range of human abilities	Wide range (10^{12}) and variety of stimuli dealt with by one unit, e.g. eye deals with relative location, movement and colour
	Insensitive to extraneous stimuli Poor pattern detection	Affected by heat, cold, noise and vibration Good pattern detection Can detect very low signals Can detect signal in high noise levels
Overload reliability	Sudden breakdown	Graceful degradation
Intelligence	None Incapable of goal switching or strategy switching without direction	Can deal with unpredicted and unpredictable Can anticipate Can adapt
Manipulative abilities	Specific	Great versatility and mobility

and machine components of the system. For example, whereas the machine may be ideal for routine repetition, the person can deal with unpredictable events better.

The Fitts List approach has at least four main disadvantages:

1 it tends to become quickly outdated with the current rate of development in areas such as microelectronics technology;
2 it can only offer a rough guide in the first stages of design and should be modified towards the final stages of process;
3 no adequate systematic methodology exists in which highly quantified engineering data can be reliably contrasted with comparable data on human performance; and
4 allocation of function should allow the individuals an opportunity not only to utilize their existing skills but also to develop.

Much of the data needed for the design of effective and safe person–machine systems are derived from systematic analyses of existing (and potential) jobs and tasks (i.e. job/task analysis).

Task analysis in system design and operation

The earlier part of this chapter stressed the need to consider person-centred approaches in the design of safe systems, this part extends the discussions (see Chapter 6 for earlier discussions) of task analysis in systems design and operation.

The design of all systems tends to pass through similar phases; from the initial design concept of the system, through its preliminary and then detailed design phases, to the systems construction, commissioning and operation, and ultimately its decommissioning. This process is known as the 'system life-cycle' (see Figure 11.4). Task analysis can be undertaken at any stage of the life cycle of a system provided that the information requirements of the particular task analysis technique can be met. For safe and effective systems design, it is cost effective to incorporate the analysis early in the design stage and, in such cases, it is also possible to establish a two-way flow of information with system users. This enables designers to incorporate the requirements outlined in earlier parts of this chapter into the proposed system and avoids the necessity for retrofit design solutions later on in the project cycle.

Task analysis is also used in other safety-related areas, including risk assessment, and in the preparation of formal job specifications. Job specifications are an important requirement for organizational selection and training (see Chapter 6).

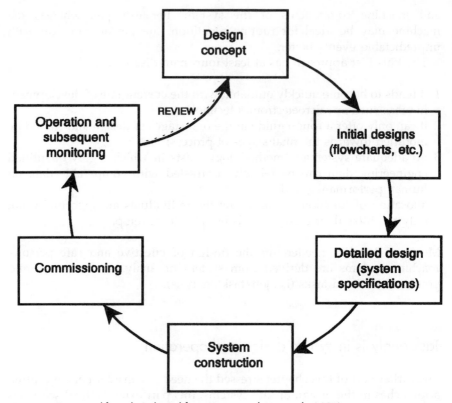

Figure 11.4 System life cycle (adapted from Kirwan and Ainsworth, 1992)

Task analysis for risk assessment

Task analysis has gained new currency in many risk assessment techniques (see Chapter 2 and S Cox, 1992). Gathering the necessary information for subsequent assessment using, for example, techniques such as observation of task completion and critical incidents, is easier in the analysis of existing jobs and tasks than for new work areas. The application of task analysis in the safety arena has also 'revitalized' techniques such as job safety analysis (see later) which are solely directed at risk control. Practical 'job' and 'task' analysis techniques, including job safety analysis, are described below.

Practical techniques for job and task analyses

The terms 'job analysis' and 'task analysis' are often used interchangeably; although it is generally accepted that a particular job may be made up of many different tasks. Job analysis usually takes one of two forms depending on the nature of the work being examined: hierarchical (top down) or sequential analysis:

1 *Hierarchical (top down) methods* force a description of the job in terms of its constituent tasks and subtasks, each level being identified and broken down into its component tasks until the integrity of these elements of work is challenged. The organization of these tasks and subtasks is represented in a tree-like structure: 'top down'. This form of analysis is applicable to most forms of work but is most helpful where the work is complex in structure and is not repetitive. An example may be provided

Table 11.5 Top down analysis of laboratory technician's job

Job	Laboratory technician
Main Tasks:	**Analysing samples** Maintaining laboratory records Stocking laboratory Disposing of waste chemicals Supervising trainees
Subtask 1:	**Carrying out lead analyses**
Subtask 1/Aspect 1	Planning analysis
Subtask 1/Aspect 1/Activities	Planning experimental design Studying written procedure Reading background information Consulting laboratory manager
Subtask 1/Aspect 2	Conducting analysis
Subtask 1/Aspect 2/Activities	Setting up equipment, etc. Completing analysis Recording results

by part of an analysis of a laboratory technician's job (see Table 11.5). (In practice the analysis would run to several pages.) The tree that may be constructed from these data is represented in Figure 11.5.

2 Simpler and more repetitive jobs may be better described by a method of *sequential analysis*. This analysis is more 'procedural' in nature. Here the main components of work are identified and the order or flow of the work described. Such data are often presented in a flow chart, and this form of analysis is useful where work is predictable and repetitive in

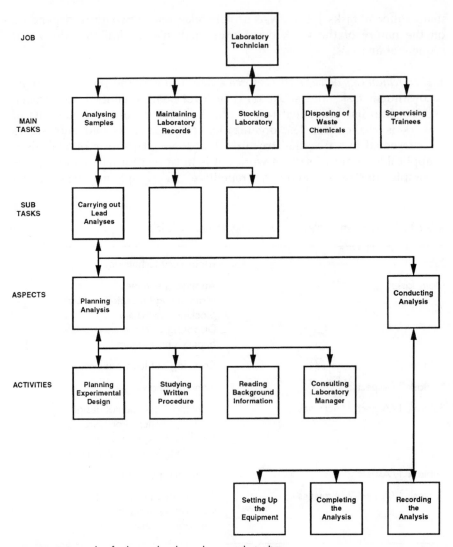

Figure 11.5 Example of a hierarchical top down analysis diagram

nature. This is represented in Figure 11.6 with reference to a train driver's sequence of actions.

In addition to these two forms of job analysis, there are other applications developed specifically for safety (Bamber, 1993), for example, job safety analysis. The first author's approach to job safety analysis is described below.

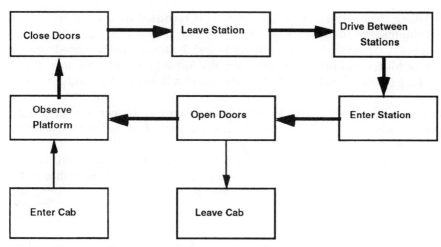

Figure 11.6 Example of a sequential analysis diagram

Job safety analysis

Job safety analysis (JSA) is a process which analyses jobs from a safety perspective. It is a method of generating potential hazards and reviewing control options within a certain job. The key aspects are described under the four main action areas of selection, team building, implementing and recording:

1 *Selecting jobs for JSA* – Jobs are usually selected using 'critical job/task criteria' (see later). For example, either (a) there is a high potential for severe injury or damage or (b) the work is new, resulting from a change in equipment, process or procedure or (c) accidents or incidents have occurred during operation, resulting in injury or unplanned events.
2 *Assembling the JSA team* – The team should include the line supervisor for the job and the job incumbent. Whenever possible trainees should participate in and study safety analyses of jobs which will form part of their work.
3 *Carrying out a JSA* – The job is first broken down into a sequence of steps, each starting with an action word such as 'remove' or 'mix', together with a description of what is being done. An experienced worker is then briefed on the purpose of the exercise and asked to do the job whilst the other participants observe and record each step.

The JSA team then review the tasks to identify any potential hazards, ask the 'what if' question, develop corrective actions to control the hazards and define a safe procedure for the job. This includes a statement of the key tasks.

4 *Filling out the JSA form* – The analysis is recorded on a form with three columns; the first shows the basic steps of the job (described by 'doing' words), the second potential hazards and the third recommended safe job procedures and other corrective actions. Table 11.6 illustrates an example of a job safety analysis for the charging of a battery on a forklift truck.

JSA may also be utilized to provide information for generic job-hazard profiles which may subsequently be used to provide data for job-capability analyses.

Essentially the same methods of investigation (data collection) can be used with hierarchical, sequential and job safety analyses. Where they differ is in the way that the results are presented.

Table 11.6 Example of Job Safety Analysis: job – forklift truck battery charging and maintenance (step 1 – battery charging) (adapted from Diploma in Health and Safety Management course notes – Loughborough University of Technology)

	Job step (task)	Hazard	Control Action
1	Park truck, secure handbrake	Unintentional movement, strain to wrist	Apply handbrake properly, avoid snatching, rapid movement
2	Switch ignition off	Electrical arc on disconnecting battery	Check ignition switch
3	Disconnect battery connector	Strain to wrist, damage to cable	Avoid snatching, rapid movement. Do not pull on cables
4	Ensure charge main switch is off	Electrical arc on connection, wrong charge allowance	Check switch. Ensure instructions relevant to the individual charger are followed
5	Check charger body, cables and connectors for damage	Fire, burns or electrocution	Report damage to supervisor, do not use charger
6	Connect charger to truck battery	Damage to connectors and nipping of hands	Do not force into place
7	Turn charger main switch on and check for charge	Insufficient charge in battery – possible truck failure later	Check amp meter or digital/LED indicator

Data collection

Just as there are special techniques which have been developed for the description of jobs and their component tasks, there are different ways of harvesting such data. These reflect a mixture of sources and methods.

Table 11.7 sets out several different ways in which job data may be collected.

Table 11.7 Job data – sources and methods of collection

From	By
Job incumbents	Observation
Supervisors	Film (video)
Managers	Interview
Workers' representatives	Group discussion
Other 'experts'	Questionnaire/video
Participant observation	Written descriptions
	Participation
	(Records)

Identifying 'critical tasks' and 'critical' personnel

Critical *tasks* may be identified by asking the following questions:

1 Can this task, if not done correctly, result in hazard potential while being performed?
2 Can this task, if not done correctly, result in hazard potential, after having been performed?
3 Is there a safe working procedure for this task?
4 Has a JSA (see earlier) been carried out recently?
5 What is the severity of harm/consequence of any incident? Is it low, medium or high?
6 What is the expected frequency (or likelihood) of an incident occurring? Is it low, medium or high?
7 Have any of the tasks been modified recently?

Most accidents and incidents occur in a work area to which a worker is unaccustomed and during a task that is not a worker's usual task (HSE, 1987). It is therefore important to remember that people are an important factor in the initial analysis and description of work activities.

Critical *personnel* may be identified using (*inter alia*) the following questions:

8   Are employees adequately trained in all health and safety aspects of the task?
9a Do all personnel routinely carry out this task or is it only carried out by certain (competent) persons? Is anybody able to cover for these people if they are unavailable either through sickness absence or illness?
9b Is anyone in the organization temporarily assigned to the task?
10 Are there any contractors or trainees in the vicinity of the task?
11 Is anyone assigned to the task suffering from any temporary or permanent disability?

Critical tasks and critical personnel (or human factors) often emerge from the Job Safety Analysis described earlier.

The working environment

Previous sections of the chapter have focused on the importance of safe design and have discussed techniques such as job and task analyses which have been used to enhance the design process. This part of the chapter briefly considers some of the features of the physical work environment and the plant and equipment (hardware) which support safe work. Many of these features are incorporated into an ergonomic survey (see Table 11.8, adapted from Rodgers *et al.*, 1986).

Six categories of descriptors are given in the survey including: workplace characteristics and accessories (equipment), physical demands, environment (physical), environment (mental), perceptual load, and displays, controls, and dials. Each category includes four to twelve specific descriptors of conditions that should be looked for in the survey. The list is intended to remind the surveyor (or designer) to look for these potential human factors or ergonomic problems, not to define the magnitude of the problem. This approach assumes that the surveyor has some experience in this type of survey and knows what to look for. It supplements other tools and techniques, for example reliability databases for plant and equipment.

The following sections highlight several issues in relation to the working environment including reliability of equipment and physical conditions. These topics are both subject to workplace regulations (see later).

Reliability of plant and equipment

One of the important considerations (particularly in the choice or design) of work equipment for safety is its reliability (see S Cox and Tait, 1991) for a

Table 11.8 Human factors/ergonomics survey descriptor list (adapted from Rodgers *et al.*, 1986, Eastman Kodak Company) (from *Reliability, Safety and Risk Management*, Cox, and Tait, 1991)

Workplace characteristics and accessories (equipment)
Reaches
Clearances
Crowding
Postures required
Chairs and footrests
Heights
Location of controls and displays
Motion efficiency
Workplace accessibility (as in moving supplies into it)

Physical demands
Heavy lifting or force exertion
Static muscle loading
Endurance requirements
Work–rest patterns
Frequency of handling
Repetitiveness
Grasping requirements
Size of articles to be handled: very large or very small
Sudden movements
Stair or ladder use
Tool use

Displays, controls, dials
Size/shape relative to viewing distance
Compatibility
Display lighting
Labelling
Internal consistency

Environment/physical
Noise level and type
Vibration level
Temperature
Humidity
Air velocity/dust and fibres
Lighting quantity
Lighting quality, especially glare
Electric shock potential
Floor characteristics, including slipperiness, slope, smoothness
Housekeeping
Hot and cold surfaces
Protective clothing needed

Environment (mental)
Skill requirements
Multiple tasks done simultaneously
Pacing
Training time needed
Monotony: low challenge
Concentration requirements
Information demands including processing
Complexity of decision making, defect recognition

Perceptual load
Visual acuity needs
Colour vision needs
Space and depth perception
Requirements
Tactile requirements
Darkroom vision
Auditory demands
Stress

detailed discussion of reliability). Over the past fifty years, particularly in safety critical systems (for example, in nuclear, aerospace, oil and gas, chemical transport and medical sectors), there has been a shift away from reliance on single components always performing satisfactorily to achieve an 'acceptable' level of reliability. More recently, the design of high reliability equipment has involved the combination of three district approaches (S Cox and Tait, 1991): quality, diversity and redundancy:

1 the quality approach to plant and equipment design invests heavily in techniques such as quality control, quality assurance and enhanced quality of components;
2 the redundancy approach to equipment design uses more than one component to perform a function, for which a single component is only strictly necessary; and
3 the diversity approach to equipment design seeks to reduce any potential problems with the work system design that could result from common mode failures (S Cox and Tait, 1991) that could arise if redundant components are not truly independent.

Brown and Hollywell (1995) have suggested that 'reliable' equipment and plant should: (1) employ high quality components (in terms of their selection and use); (2) have some redundancy of function; and (3) have some degree of diversity to overcome the potential reliability problems that can come with using redundant components (see S Cox and Tait, 1991). They further suggest that such principles can be readily translated into designing for human reliability (Hollywell, 1994).

Physical conditions

The importance of the physical conditions in work environments (see Table 11.8) should not be understated. The regulations on workplace health, safety and welfare (HSC, 1992) require employers to address 'workplace' hazards including, for example, workplace layout (room dimensions, space and workstation design), flooring, ventilation and thermal environments, lighting, etc. Specific workplace regulations on noise supplement these requirements in relation to the levels of ambient noise. Employers also have a general duty under Section 2 of the Health and Safety at Work etc. Act 1974 (HASAWA) to ensure, so far as is reasonably practicable, the health, safety and welfare of their employees at work. They should ensure that workplaces under their control comply with all these regulations.

The physical conditions in the workplace support safe work and there is a requirement for employers to adequately assess and control the risks associated with such hazards (see Chapter 2). However, it is also important to ensure that the physical environment is routinely monitored and reviewed as part of an ongoing maintenance of the safety system. Audit protocols (see earlier) can support this process. The following sections briefly consider safety issues in relation to workstations and work equipment.

Workstations

Workstations should be arranged so that each task can be carried out safely and comfortably. The worker should be at a suitable height in relation to the work surface. Work materials and frequently used equipment or controls should be within easy reach, without undue bending or stretching. Each workstation should also allow any person who is likely to work there adequate freedom of movement and to be free from obstructions or clutter.

One of the most frequently reported risks to workers in the UK is that of slipping, tripping and falling (see S Cox and O'Sullivan, 1995). Workplace assessments should recognize such risks and consider the presence of potential slipping and tripping hazards including uneven or slippery floor surfaces, slopes, any potential leaks or spillages, stairs, etc.

Work equipment

The Provision and Use of Work Equipment Regulations 1992 (PUWER) lay down important health and safety laws for the provision and use of work equipment. The primary objective of PUWER is to ensure the provision of safe work equipment *and* its safe use. This has several components including (*inter alia*):

1 the exact nature of the work equipment, whether it is new or secondhand, is already in use or is hired or leased;
2 its suitability for the job;
3 its maintainability;
4 specific risks, for example those associated with dangerous parts of machinery;
5 energy sources; and
6 operator training and information and instructions.

The reader is referred to the guidance booklet on PUWER (HSE, 1992) for a more detailed account of these factors. One of the key factors in the provision and safe use of work equipment relates to the provision of appropriate information to the user. This may be achieved in a number of ways, for example, through training and the 'use' of written procedures. If working procedures are used they should be made clear and unambiguous, and be prepared in a user friendly manner at the right level of knowledge and understanding for the recipient. Working procedures should be compatible with safety rules and, where necessary, permit-to-work systems should be used. Feedback on working procedures should be actively encouraged as part of the safety review process.

Patterns of work: shift systems

There are several important issues in relation to the organization of work and working hours. These include the length and regularity of the work period, the balance between work and rest during these periods and the major question of shiftwork. Many of the problems underpinning these three issues are the same and most are represented in discussions of shift working. The focus of this brief review is therefore the whole question of shiftwork.

There are many types of shift systems in use in industry both fixed, i.e. the employee works the same hours throughout their working life, or rotating. Rotating shift systems may be subdivided into those systems with and without night work and systems with and without work on Sundays. Knauth (1993) has summarized other characteristics which characterize shift systems, such as the number of consecutive shifts, the direction of the rotation and the distribution of leisure time. These are included in Table 11.9.

Table 11.9 Additional features which characterize shift systems (adapted from Knauth, 1993)

1	Number of consecutive	– morning shifts
		– evening shifts
		– night shifts
		– working days
2	Length of each shift	
3	Times at which shifts start and finish	
4	Distribution of leisure time	– the duration of time off between two shifts
		– the time of day at which leisure time is available
		– the time of week at which leisure time is available
		– periodic components
5	Direction of rotation	
6	Regularity of shift system	
7	Flexibility of shift system	
8	Part-time work, full-time work	

There is no single 'optimum shift system' which can be applied to all work organizations. Factors such as local collective agreements, the state of the labour market and associated economic conditions, the nature of the task (for example, whether the environmental conditions argue for increased job rotation) and finally the prevailing culture will all be relevant.

Knauth (1993) has reviewed the available scientific literature on the design of shift systems in industry and commerce and, although he suggests that all shift systems have both advantages and drawbacks, he maintains that there are some shift systems which are more favourable than others. His recommendations are made on the basis of physiological, psychological and social factors and include the following:

1 Night work should be reduced as much as possible, although if this should not prove possible, then quickly rotating systems are preferable to slowly rotating ones.
2 Extended workdays (nine to twelve hours) should only be contemplated if the nature of the work and workload are suitable. Knauth (1993) argues that the shift system is designed to minimize the accumulation of fatigue. Organizations should also ensure that adequate cover is available for absentees.
3 An early start for the morning shift should be avoided where possible. In all shift systems flexible working time arrangements are realisable, the highest flexibility being achieved in 'time autonomous groups' (ILO, 1990).
4 Quick changeovers (for example, from night to afternoon shift on the same day or from afternoon to morning shift) should be avoided (Folkard and Monk, 1985). The number of consecutive working days should be limited to five to seven days. Every shift system should include some free weekends with at least two successive full days off.

Shift systems and safety

There is a general consensus (Waterhouse et al., 1992) that productivity and safety may be reduced on the night shift, but the empirical data is equivocal. Errors appear to be up and general performance down at night, but if the work is designed so as to include regular breaks and repetitive and boring tasks are minimized then the night-shift worker should not be worse off than their daytime colleagues.

Interpreting reductions in productivity and safety at night is further complicated by the fact that conditions differ between daytime and night-time working. For example, communication systems may differ, manning levels may be generally lower (for example, maintenance workers may not work night shifts) and ambient lighting levels may be lower. Despite this, Waterhouse et al. (1992) have identified the following issues as important for safety:

1 rhythmic changes – our circadian rhythms, which are linked to our levels of physiological arousal, are normally at a low ebb in the small hours;
2 individuals are more fatigued and more likely to fall asleep on the night shift than during the day; and
3 shifts over eight hours tend to be associated with declining performance if the task is physically or mentally demanding or repetitive. Long shifts may thus contribute to accidents; however, the nature of the link between poorer performance and accidents remains to be established.

Waterhouse *et al.* (1992) conclude 'There is a growing body of evidence which indicates that the efficiency of the workforce and health of some of its members may be affected by problems associated with shiftwork'. However, many studies have suffered from lack of standardization of methods and the variables that have been measured in them. In relation to this text, however, shift pattern and fatigue are important considerations of the job and need to be optimized in the light of particular organizational systems and procedures.

In summary, this section of the chapter has examined some of the issues underpinning the organization of work, particularly in relation to shiftwork. It has summarized some of the evidence linking shiftwork and accidents. The final section of this chapter examines several accidents and incident scenarios to illustrate some of the consequences of poor design.

Learning by example: the pathway to safer designs

The safety literature is full of examples of 'poor' or inadequate design of jobs and tasks which in turn have created unacceptable risks to workers. This section illustrates some recent examples. It thus also demonstrates the importance of organizational learning discussed in the previous chapter (see Figure 10.9). The authors attempt this at two levels – first, in relation to everyday industrial and commercial processes, and second, in relation to major disasters.

Examples of poor design in everyday processes

The first example concerns an agricultural task (although similar examples can be found in the construction industry) and it is quoted directly from the HSE publication *Human Factors in Industrial Safety* (HSE, 1989).

> An experienced driver was driving a tractor along a farm road. He was using the tractor to haul a hydraulically operated tip-up trailer loaded with grain. The gear lever in the tractor cab was positioned close to and in line with the hydraulic control lever which operated the tip-up mechanism of the trailer. In pulling the gear lever backwards, the driver's arm moved the control lever slightly so that the trailer began to move slowly upwards without the driver's knowledge. The rising trailer eventually reached a position where it affected the stability of the tractor which subsequently overturned (HSE, 1989, p.4).

The second example demonstrates the importance of worker expectations (or stereotypes) to work-equipment. Graves and Pethick (1987) describe a classic example of poor press design where the operator was required to *raise* a lever to bring down the press and to *lower* the lever to lift the press.

Wishing to raise the press in an emergency, the operator lifted the lever up and the platen moved down and the press was wrecked. Whereas the operator seemed to be coping with a 'non-stereotypical' design feature during normal operating conditions, during an emergency situation he reverted to design stereotypes. Several authors (see, for example, Rodgers *et al.*, 1986) have described some of the common stereotypes that need to be considered in safe designs.

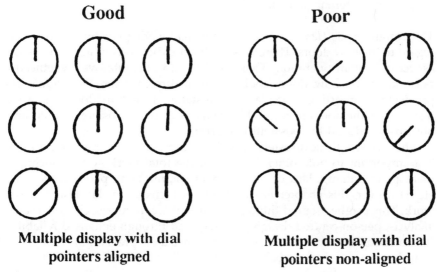

Multiple display with dial pointers aligned

Multiple display with dial pointers non-aligned

Figure 11.7 Alignment of dials to detect normal functioning (adapted from Rodgers *et al.*, 1986)

The third example (Rodgers *et al.*, 1986) illustrates the importance of 'good' instrument design for safe practices (see Figure 11.7). In Figure 11.7, two examples of dials located adjacent to each other on a display panel are shown. In the good example (a) all the 'pointers' are aligned similarly for normal functioning. In this case the operator can scan the displays easily and detect any plant changes. In the poor example (b) the pointers are not aligned and plant monitoring is particularly difficult.

Inadequate instrument design has been implicated in the causation of several major accidents.

Examples of poor design in relation to major disasters

Two of the more spectacular examples of inadequate consideration of ergonomic and psychological factors in systems design are the Three Mile Island Nuclear Plant Incident (Rubenstein and Mason, 1979) and the *Herald of Free Enterprise* ferry accident (Department of Transport, 1987).

The official report on the Three Mile Island power station cited the following examples of bad design:

1 Poor control room design in which there were examples of mirror image layouts on duplicate panels and poorly positioned dials.
2 Poorly designed display systems. The operators had provided their own improvised solutions by relating the relevant display and control together by arrowed sticky tape. In a second example two identical control knobs had very different functions.

The captain of the *Herald of Free Enterprise* had inadequate information with regard to the bow doors. There was no direct indication on the bridge of this aspect of system operation. There is also some dispute over whether the position of the bow doors could be seen directly from the bridge. The provision of an artificial display on the status of the bow doors and a real display (i.e. the doors being visible from a reasonable position) would have obviously reduced the possibility of error.

Whereas both these disasters provide examples of poor ergonomic design, it is important to note other features of the total work system, including inadequate training, lack of management commitment, poor maintenance, etc. However, this chapter has particularly focused on the design of safe work. This can be assured through the use of techniques such as job safety analysis, person-centred design processes and through enhanced reliability. It also requires a systematic approach to accident prevention through the consideration of error probabilities as described in Chapter 6. There is no simple panacea for the prevention of accidents such as those described earlier in this section. What is required is a concerted systems approach which considers jobs and tasks in their wider context. The physical working environment is only part of the system but it is particularly important in shaping workers' perceptions of the organization's commitment to safety. Common sense tells us and research confirms that workers use physical appearances to make judgements about safety commitment. This is further explored in the next chapter.

Summary

This chapter has considered the principles underpinning the design of safe work in the light of current psychological and ergonomic knowledge. It has particularly stressed the importance of the physical environment in shaping workers' approaches to safety and has explored the issues surrounding work organization with reference to shift systems. It has also demonstrated the use of job/task analysis and the design of safe work and has finally explored how poor design can contribute to unsafe systems.

References

Bamber, L. (1993). 'Techniques of accident prevention'. In *Safety at Work* (J. Ridley, ed), Butterworth-Heinemann.

Boff, K.R. and Lincoln, J.E. (1988). *Engineering Data Compendium, Human Perception and Perception.* Wright-Patterson Air Force Base, OH 45433, USA: H.G. Armstrong Aerospace Medical Research Laboratory, Human Engineering Division.

Brown, M.R. and Holywell, P.D. (1995). 'Human reliability in safety critical systems'. In *Proceedings of Ergonomics Society Conference, April,* 499–504.

Cox, S. (1992). *Risk Assessment Toolkit.* Loughborough: Loughborough University of Technology.

Cox, S. and O'Sullivan, E. (1995). *Building Regulations and Safety.* Watford: Building Research Establishment.

Cox, S.J. and Tait, N.R.S. (1991). *Reliability, Safety and Risk Management.* London: Butterworth-Heinemann.

Cox, T. (1985). 'Repetitive work: occupational stress and health'. In *Job Stress and Blue Collar Work* (C.L. Cooper and M.J. Smith, eds). Chichester: Wiley and Sons.

Cox, T. and Cox, S. (1993). *Psychosocial and Organisational Hazards at Work: Control and Monitoring.* European Occupational Health Series 5 (Europe: World Health Organization).

Department of Transport (1987). *M.V. Herald of Free Enterprise: Report of the Court No. 8074* (Department of Transport Report, 1987). London: HMSO.

Fitts, P.M. (1951) *Handbook of Experimental Psychology.* Chapter on Engineering Psychology and Equipment Design, London: John Wiley.

Folkard, S. and Monk, T.H. (1985). *Hours of Work.* Chichester: John Wiley.

Framework Directive 89/391/EEC (1989). *The Minimum Requirements for Protecting the Health and Safety of Workers.* Brussels, 12 June 1989.

Graves, R.J. and Pethick, J. (1987). 'Designing for occupational health and safety'. *The Safety Practitioner,* December, 10–15.

HAZDATA (1990). National Chemical Emergency Centre, UKAEA. Harwell, Oxford, UK.

Health and Safety Commission (HSC) (1992). *Workplace Health, Safety and Welfare.* London: HMSO.

Health and Safety Executive (HSE) (1987). *Dangerous Maintenance in the Chemical Industry.* London: HMSO.

Health and Safety Executive (1989). *Human Factors in Industrial Safety.* London: HMSO.

Health and Safety Executive (1992). *Work Equipment Guidance on Regulations.* London: HMSO.

Hollywell, P.D. (1994). Incorporating Human Dependent Failures in Risk Assessment to Improve the Estimation of Actual Risk. *Proceedings of Risk Assessment and Risk Conference,* Aston University, 22 March, Aston.

International Labour Office (ILO) (1990). 'Conditions of work design – the hours we work: new work schedules'. *Policy and Practice,* 9(2).

Knauth, P. (1993). 'The design of shift systems'. *Ergonomics,* 36(1–3), 15–28.

McCormick, E.J. (1983). *Human Factors in Engineering and Design,* 4th edn. USA: McGraw-Hill.

Meister, D. (1989). *Behavioural Foundations of System Development*. London: John Wiley.

MHIDAS Safety and Reliability Directorate, Wigshaw Lane, Culcheth, Warrington, UK.

Oborne, D.J. (1995). *Ergonomics at Work*, 3rd edn. London: John Wiley.

Rodgers, S.H., Kenworthy, D.A. and Eggleton, E.M. (1986). *Ergonomic Design for People at Work*. New York: Eastman Kodak Company, Van Nostrand Reinhold.

Rubenstein, T. and Mason, A.F. (1979). 'The accident that shouldn't have happened: an analysis of Three Mile Island'. *IEEE, Spectrum*, November, 35–37.

Sen, T.K., Pruzansky, S. and Carroll, J.D. (1981). 'Relationship of perceived stress to job satisfaction'. In *Machine Pacing and Occupation Stress* (G. Salvendy and M.J. Smith, eds) pp. 65–71, London: Taylor and Francis.

Singleton, W.T. (1974). *Man Machine Systems*. Harmondsworth: Penguin Books.

Waterhouse, J.M., Folkard, S. and Minors, D.S. (1992). *Shiftwork, Health and Safety –An Overview of the Scientific Literature 1978–1990*. HSE Contract Research Report No. 31.

12
Managing people and their attitudes to safety

Introduction

Previous chapters in this section have considered frameworks for the management of safety, the necessary organizational systems and issues related to job design. This chapter focuses on individuals and the attitudes necessary for their effective involvement in safe systems of work. It discusses how these might be developed and managed in the workplace.

The practice of safety management is often said to be about 'winning hearts and minds'. While management is quite obviously more than this (see Chapter 9), the phrase does capture a number of its key elements. 'Winning' implies a challenge for the managers involved and the need to change existing situations. It stimulates thinking about the different techniques which may be required to bring about those changes, such as persuasion, negotiation and inducement. The term 'hearts and minds' recognizes the importance of emotion and cognition as the psychological processes which drive behaviour and behaviour change. Combining the notions of 'winning' with that of 'hearts and minds' focuses the practice of management on the challenge of, and techniques required for, changing existing cognitions and emotions as a prerequisite for changing behaviour. Here the authors explore the application of this simple model to managing safety; much of it overlays the literature on attitudes to safety and related attitude change (see later). Richardson (1977), for example, has offered a model of attitudes to safety which links cognitions and emotions (affect) to behaviour in the *demonstration* of an attitude. While it is still not clear *how* exactly the former, cognitions and emotions, relate to behaviour, it is obvious that the three *are* related (see Ajzen, 1991) and change in one may be associated with change in one or both of the others. Under some circumstances, this observation may provide a basis for intervention and for planning how to 'win hearts and minds' for safety (see Figure 12.1).

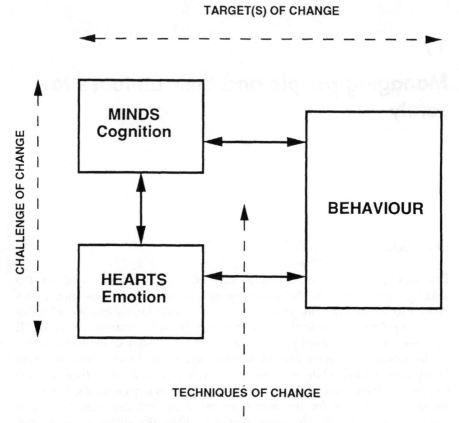

Figure 12.1 The concept of winning hearts and minds

This chapter presents a particular view of attitudes (see also Chapter 7) and the processes underpinning attitude change for safety (the target of change). It considers the necessary skills of communication, persuasion and negotiation (the techniques of change). It then considers the relationships between attitudes and behaviour, and discusses the first author's approach to behaviour based safety programmes. It finally considers how training may be used to facilitate the process of 'winning hearts and minds' so that safety-positive attitudes and safe behaviour can be developed and supported.

Attitudes

Both social and occupational psychologists have been concerned to define the nature of the relationship between a person's attitudes (see Chapter 7) and their behaviour. Although it would seem logical that attitudes, as frameworks for decision making, should be lawfully related to behaviour, that relationship has proved complex and elusive, and its 'rules' have proven challenging to discover. However, because attitude change seems to hold the promise of sustained behavioural change, psychologists have persevered with this particular quest. The basic questions are: (1) what is the relationship between attitudes and behaviour; (2) how can attitudes be changed and, in the present context; (3) is attitude change important for the management of safety? These questions are addressed below.

Attitudes and behaviour

There are a number of theoretical approaches to the possible relationship between attitudes and behaviour. Glendon and McKenna (1995) have articulated four possible types of relationship (see Figure 12.2). These are:

1 Attitudes influencing behaviour – if we know a person's attitude to some safety-related practice (for example, wearing safety spectacles) then we can predict their relevant behaviour, or if we wish to change that behaviour we can achieve this, in large part, by changing the associated attitude.
2 Behaviour influencing attitude – if we want to change someone's attitude towards wearing safety spectacles, we can achieve this by obliging them to behave in a particular way, for example by making and enforcing a rule about wearing safety spectacles in certain work areas.
3 Attitudes and behaviour influencing each other. If we are able to change either then the other will change accordingly; attitudes and behaviours will achieve congruity. The notion of congruity or consistency underpins Festinger's (1957) theory of cognitive dissonance which states that people strive for consistency in attitudes and behaviour.
4 Other factors (for example, organizational campaigns) are involved in the attitude-behaviour equation and may influence both attitudes and behaviours, and do so independently.

Not all of these logically possible relationships are supported by the empirical evidence. Several studies have, for example, found attitudes to be poor predictors of behaviour (see, for example, Glendon and Hale, 1984). Furthermore, it has also been shown that changes in attitudes are not necessarily translated into corresponding changes in behaviour. On the other hand, legislation in seatbelt usage has been successful in changing behaviours related to seatbelt wearing and, in line with cognitive dissonance theory (Festinger, 1957), attitudes appear to have changed to remain consistent with the newly acquired behaviour (see Kleinke, 1984). It would seem, on the balance of the evidence, that there can be some

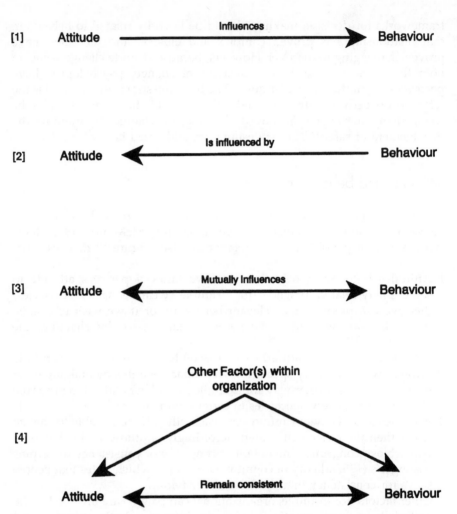

Figure 12.2 Possible linkages between attitudes and behaviour (adapted from Glendon and McKenna, 1995)

correspondence between the attitudes and behaviour but that that may hold true 'for some individuals more than others' and 'for some situations more than others' (Rajecki, 1990). A possible 'person × situation' interaction is thus indicated and should be part of any explanation of the relationship between attitudes and behaviour. Snyder and Kendzierski (1982) have offered such an explanation in terms of the conditions necessary to allow attitudes to be translated into behaviour. Their ideas have been utilized in the first author's approach to attitude change and are described in the following section.

Snyder and Kendzierski's model

Snyder and Kendzierski (1982) have outlined the two conditions that must be met before a person's attitudes can be translated into behaviour: *availability* and *relevance*.

The first condition, *availability*, refers to the availability of personal knowledge about one's attitudes. Before a person can use their attitudes as a guide to action, they must know what they are. Accordingly, they must define an attitudinal domain applicable to the specific situation and behavioural choices that confront them, and bring to mind the beliefs, feelings and behaviour associated with that domain. Snyder and Kendzierski (1982) argue that availability of knowledge about attitudes may nevertheless be insufficient on its own to guarantee correspondence between attitudes and behaviour. The second condition then comes into operation: *relevance*. Before the person can act on their knowledge, that knowledge must be defined as relevant to their actions. They must (1) regard their attitudes as having important implications for their actions, and (2) believe that, therefore, their attitudes ought to be related meaningfully to their behaviour.

There may be a variety of different reasons why attitudes might not be defined as relevant to behaviour; other situational and social factors may be believed to have greater importance in determining behaviour, for example, peer group pressure, role requirements, management incentives or sanctions. These other factors may reflect safety culture (Lee, 1993), possibly more directly than attitudes do. Such factors may also lead the person to behave in such a way as to appear to contradict their own attitudes.

Snyder and Kendzierski (1982), while recognizing that availability and relevance are both necessary conditions for correspondence between attitudes and behaviour, suggest that relevance may be a sufficient requirement. Redefining one's attitudes as relevant presupposes, and indeed depends, on a knowledge of what those attitudes are. They further suggest that there are significant individual differences in the way the relevance condition is handled. From their research, there would appear to be some people who define available knowledge of attitudes as behaviourally relevant knowledge, while others tailor their behaviour to meet situational or social requirements. The former group have been identified by their low scores on Snyder's Self-Monitoring Scale (1979), and the latter by their high scores. Low self-monitoring individuals reportedly endorse statements such as:

1 'my behaviour is usually an expression of my true inner feelings, attitudes and beliefs'; and
2 'I can only argue for ideas which I already believe'.

The availability of knowledge of their attitudes is usually sufficient to prompt this group to translate that knowledge into behaviour which meaningfully reflects those attitudes. By contrast, high self-monitoring individuals apparently endorse statements such as:

1 'I sometimes appear to others to be experiencing deeper emotions than I actually am'; and
2 'I am not always the person I appear to be'.

For these individuals, available knowledge of attitudes may not be sufficient to guarantee correspondence between attitudes and behaviour. They require situations or social roles which provide clear specifications that their attitudes ought to be relevant to behaviour in order for that correspondence to be achieved. Snyder and Kendzierski (1982) have suggested that attitude change strategies based on manipulations of declared relevance will effectively enhance correspondence between attitudes and behaviour to the extent that they induce people to invest in 'believing means doing'. Snyder and Kendzierski's (1982) model has implications for training strategies (see later). However, while it recognizes the importance of individual processes for behavioural outcomes, it does not address in sufficient detail the nature and role of situational factors. These are considered by Ajzen (1991) in his theory of planned behaviour.

Theory of planned behaviour

The theory of planned behaviour (Ajzen, 1991) suggests that behaviour is driven by the *intention* to act (see Figure 12.3). 'Intentions' (at both conscious and unconscious levels) are shaped by three factors:

1 the person's prevailing attitude towards the behaviour;
2 perceived social norms, which are derived from immediate groups; and
3 perceived behavioural control.

For example, a person's intention to perform a particular act, such as wearing hearing defenders, is not *only* driven by the person's attitude towards the subject of 'noise control' but it combines with other social and motivational factors. Ajzen's theory thus addresses the role of situational factors for behavioural outcomes.

For safety, the final factor can be related to the person's perceptions of hazard and risk (see Chapter 7) and their beliefs in their ability to control such hazards. Perceived behavioural control also feeds directly into the behavioural outcome (Ajzen, 1991). For example, car drivers may express safe attitudes to driving and then may choose to drive at seventy miles per hour over a humpbacked bridge. Their 'planned' behaviour incorporates the

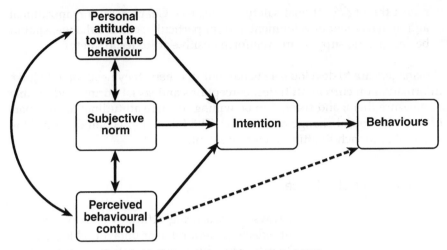

Figure 12.3 Theory of planned behaviour (adapted from Ajzen, 1991)

'misperception' that they are able to fully 'control' the car. The practical manifestation of this plan (the actual driving behaviour) will probably confirm that they are unable to do so.

The theory of planned behaviour, as suggested by Ajzen (1991), taken together with Snyder and Kendzierski's 1982 model, would further suggest that the following need to be taken into account in the process of attitude change for safety:

1 The person needs to hold an attitude to the particular aspect of safety being considered.
2 They need to appreciate the relevance of that attitude to their behaviour. For example, the worker must realize that a positive attitude to eye protection equipment has relevance if they are working alongside a compressed air line (education and training (see later) should be designed to help the development of positive attitudes, and to point out the conditions of their relevance).
3 The person needs to believe in their ability to control their own behaviour (education and training should be designed to teach people that they can control their behaviour with regards to safety). These beliefs need to match actual control.
4 The organizational context and prevailing social environment should create social norms which support and reinforce positive attitudes. For example, if a work group has social norms which value safety, then the individual will be encouraged to hold positive views. Conversely, if safety is not valued by the immediate work group, then it is unlikely to be valued by an individual worker. The social norms of the work group will usually

reflect the organizational safety culture (see Chapter 10) (organizational and management development and, in particular, culture change, should be designed to support and reinforce positive attitudes to safety).

If managers are to develop safe behaviour in others, they need, first, to have information on current attitudes, perceptions and social norms, and be able to measure them, and second, to be willing to share attitudinal data as part of the behavioural change programme. The following section describes the authors' approach to attitude measurement.

Measurement of attitudes

One of the most common ways of measuring attitudes is to present the respondent with a statement reflecting favourable or unfavourable attitudes and ask them to indicate the extent to which they 'agree' or 'disagree' with those attitudes (see Appendix 2 for an example). Fine grades of response may be permitted by including 'strongly agree' and 'strongly disagree' categories, and possibly a 'neither agree nor disagree' category. In this way, five-point response scales can be provided. This type of scale was first suggested by Likert (1932).

Scales developed using this strategy, like all other scales, require evaluation before use. Such evaluation should reveal their inherent weaknesses. One major problem concerns the reasons why a respondent might disagree with an item. This may occur for two reasons: because they hold (1) a negative attitude or (2) a strong positive attitude that the statement is not strong enough. For example, consider the statement 'employees should be reprimanded if they flout safety procedures'. A respondent could disagree with this statement because they believe (1) employees should not be reprimanded for this behaviour, or (2) that they should be dismissed for this behaviour. The scale cannot easily deal with the second possibility; it assumes that all its items are monotonic. However, this problem can be overcome, in practice, by careful initial research into the substance and strength of the attitudes under consideration.

Table 12.1 illustrates the five stages in the authors' approach to attitude measurement using Likert scaling. It has been used to model attitudes to safety in a variety of organizations including manufacturing, process industry, service and healthcare industry (S Cox and T Cox, 1991; Cheyne and S Cox, 1994). A somewhat similar approach has been used by several other practitioners in this area (see, for example, Lee, 1993 for a study of attitudes to safety within the nuclear industry and HSE, 1993 for a report on attitudes towards noise as an occupational hazard). The approach is fully explicated in relation to a study of employee attitudes to safety in a European industrial gas company (S Cox and T Cox, 1991).

The structure of employee attitudes to safety: a European example

The study was concerned with the common architecture of attitudes across occupational group and country. It used a framework described by Purdham (1984) which focuses on two main issues in relation to the architecture of attitudes to safety: the objects of attitudes to safety (including (1) safety hardware and physical hazards; (2) safety software and concepts; (3) people and (4) risk) and the subject of these attitudes (whose attitudes are they?). This was consistent with the company's safety philosophy and focused on attitudes to safety systems (software), people and risk. It did not cover employee attitudes to safety hardware or specific hazards.

Table 12.1 Attitude measurement: a five-stage process

Stage	Process
1 Initial discussion framing concerns and planning design	Focus groups or representative discussions yielding 'verbatim' records. Developing attitude statements and pilot questionnaire instrument
2 Pilot study/development	Pilot questionnaire distributed, reliability studies and subsequent refinement and validation
3 Questionnaire distribution and data collection	Refined questionnaire distributed to test population and 'confidential' data collection
4 Data analysis	Data coding and analysis using computer-based statistical packages (BMDP or SPSS). Factor analysis to explore underlying structure
5 Feedback	Feedback takes several forms including written, verbal and formal presentations

The study (S Cox and T Cox, 1991) was set within the context of a larger programme concerned with the development of the company's safety culture (see Chapter 10). The programme adopted began with an analysis of the company philosophy and existing culture. This analysis involved the development and execution of an employee attitude survey, which included a study of the reliability of the survey questionnaire. An intervention to improve attitudes to safety was then planned on the basis of the survey data, implemented and evaluated (S Cox, 1988).

Developing the attitude survey
The design and conduct of the attitude survey was subject to recognized good practice (see Table 12.1) and was comprised of a number of discrete phases including: (1) the initial discussions and design strategy; (2) the development of the survey instrument and (3) the reliability study – leading to the formulation of an acceptably reliable questionnaire.

The final instrument comprised four sections (A through D; see Table 12.2). The first two (A and B) took the form of scales for the assessment of employee attitudes. Section A considered their attitudes to good safety practices, while section B considered employee's attitudes to the company's safety philosophy and culture.

Table 12.2 Questionnaire design (adapted from S Cox, 1988)

Section	Items	No.
A	Attitudes to good safety practices	12
B	Attitudes to the company's safety philosophy and culture	6
C	Perceptions of the company's commitment to safety Attendance at 'family safety days'	3
D	Suggestions for improving attitudes to safety	–

Employees were presented with a series of statements, in their own language, about safety and the company's practices and asked to what extent they agreed or disagreed with each statement using Likert (1932) scales. Checks were made in the development of these scales to increase the probability that the scales were monotonic. Both positive and negative statements were included in the questionnaire. Five-point scales were used to record their responses (strongly agree, agree, no opinion, disagree, strongly disagree). An additional category, 'I do not understand this statement', allowed respondents to indicate where such a lack of under-standing existed. A mixture of positive and negative safety statements were presented in the two sections. The scores from the negatively phrased questions were adjusted to ensure consistency in the direction of scoring.

The third section, C, comprised a series of questions designed to assess employees' perceptions of the company's commitment to safety. The final section, D, asked for suggestions on how attitudes to safety might be improved.

Questionnaire distribution and data collection

The questionnaires were distributed to the company's employees through its various European depots, following an announcement and description of the study published in the company's journal. It was agreed that central training staff should first introduce the objectives and nature of the study to the depot management, who would then brief their safety committees and other staff. The cooperation of employees was also sought, where possible, through the vehicle of depot training days. However, this procedure was not followed in all depots.

The actual method of distribution was negotiated at local level and varied from depot to depot to take into account differences in management practices, and wider cultural differences in employee's reactions to questionnaire surveys. In the UK, Germany and Belgium, the questionnaires were directly distributed to employees by management. In France they were sent by post to employee's homes, and in Holland they were distributed by training staff. In both France and Germany, the General Managers posted letters on all noticeboards encouraging employees to complete their questionnaire.

The completed questionnaires were returned by post to the first author. It was stressed throughout that the questionnaire returns were anonymous, and confidential to the study.

Data analysis

The quantitative data, from sections A through C, were coded and then analysed using a computer-based statistical package (BMDP). The data from the main survey, which measured employee attitudes to safety in general, and to the company's safety philosophy and culture in particular (sections A and B), were subjected to factor analysis to explore their underlying structure (Ferguson and T Cox, 1993).

Possible architecture of employee attitudes

The data collected from 630 respondents within this European company described five dimensions or factors underpinning their attitudes to safety. Together these factors accounted for 50 per cent of the data variance. These five factors related to safety software (effectiveness of arrangements for safety), people (individual responsibility and personal scepticism), and risk (safeness of the work environment, and personal immunity).

These data have proved useful for three reasons. First, they extend our knowledge of the architecture of employee attitudes to safety: thus the data are of theoretical importance. On consideration of the results of the factor analysis, a tentative model can be described (see Figure 12.4). This model emphasizes the shared aspects of employee attitudes to safety (ACSNI, 1993) and to this extent provides a partial description of the company's safety culture.

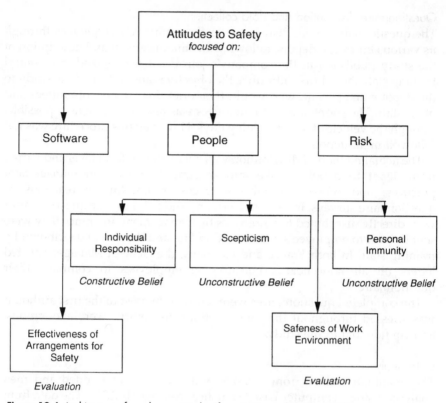

Figure 12.4 Architecture of employee attitudes (from S Cox and T Cox, 1991)

This model naturally elaborates the general framework first suggested by Purdham (1984) in his description of possible objects of safety attitudes. Although attitudes to safety hardware and specific hazards were not included in the present study, it is suggested here that these attitudes might influence (and interact) with those representing the three other objects. Certainly, the attitudes under study can be grouped in a way consistent with Purdham's (1984) distinction between those three objects: software, people and risk. Furthermore, it also suggests that, at least within this one company, attitudes about people (in relation to safety) encompass an apparently paradoxical combination of factors relating to individual responsibility and personal scepticism (about the importance of safety). Somewhat similarly, attitudes to risk reflect both the safeness of the work environment and the idea of personal immunity (through experience and expertise). It would appear to the authors that there is a mixture here of evaluation, and constructive and unconstructive beliefs about safety. This conclusion about the common architecture of employee attitudes should not be read to mean that individuals necessarily hold paradoxical beliefs about safety, although this is possible. The question at the individual level is an empirical one.

Second, building on this model, it is suggested that strategies for enhancing an organization's safety culture through attitude change (however achieved) should consider both the reinforcement of constructive beliefs and positive evaluations, and simultaneously the extinction of unconstructive and negative beliefs. Such a strategy of attitude change would parallel the behaviour modification approach described by Burkhardt and Schneider (1987) and others (see later sections). In the present case this would involve rewarding any expression of individual responsibility for safety, and building on employee's knowledge and positive evaluations of the arrangements for safety and the safeness of the work environment. At the same time, steps should be taken to change unconstructive beliefs about personal immunity, and to reduce any scepticism over safety. Indeed, this strategy was used by the first author in the subsequent intervention to enhance attitudes to safety with key groups of supervisors (S Cox, 1988). In doing so, particular attention was paid to Snyder and Kendzierski's (1982) notion of promoting 'relevance' and the belief that 'believing means doing'.

Last, the model also suggests a possible measurement or audit instrument. The first author has collected further data, using the attitude questionnaire, in several other large multinational companies (Cheyne and S Cox, 1994). These data are being used in the further development of an audit instrument (see Chapter 10).

It is generally accepted that attitudes towards 'safety' and 'safe behaviours' are one of the basic components of safety culture (see Chapter 5) and, by some mechanism, that they are both cause and effect. One of the biggest questions for people management, therefore, is how to manipulate this process in order to produce a change for the better.

Changing attitudes

There is much literature on attitude change, most of which starts with an acceptance of the commonplace experience that attitudes are notoriously hard to change. The reasons underpinning their enduring nature and resistance to change are difficult to ascertain, and require study in their own right. Writers have treated the question of attitude change in different ways. Two distinct approaches may be discerned: first, that which treats attitude change as an act of communication, and which has given rise in the present context to an investment in safety communication (including safety campaigns) and, second, that which bases attitude change on the manipulation of behaviour, for example by general laws or specific behavioural programmes. In Glendon and McKenna's (1995) terms, the latter approach explicitly draws on the position that attitudes are, to some degree, influenced by behaviour. The former, in practice, also tends to this position but usually in terms of the verbal or subvocal behaviour involved in communication. Approaches to attitude change, based on communication methodology, appear to deal in terms of three or possibly four variables: the *source of the argument* for change, the *argument* itself, and the *means by which it is delivered*. Some also refer to the *characteristics of the recipient*. The model underpinning much of this type of research assumes that attitude change, an act of verbal or visual communication, is the process of 'argument'.

Before considering *how* attitudes are changed, it is worth considering *why* a person changes their attitudes; the literature largely deals with the 'how' (how one person changes another's attitudes) at the expense of the 'why' (why a person would change their own attitudes). Kelman (1961) recognizes three reasons why a person might change their attitudes: compliance, identification and internalization. Compliance arises when an attitude is adopted for ulterior motives, such as the desire to make a favourable impression on significant others. Identification arises when the individual adopts an attitude in order to establish or maintain a satisfying relationship with others or with a particular group. Internalization arises when the new attitude is embraced as part of a cluster of attitudes because the individual feels comfortable subscribing to them. All three reasons can be explained in terms of reward mechanisms. Some form of position reward may have to be forthcoming for the person to change their attitudes, or, at the very least, the change should reduce some existing level of discomfort or dissatisfaction. As a means of encouraging workers to use personal protective equipment, changes could be made to both their design and the material used to improve the comfort of the wearer.

Safety communication

Communication, as a vehicle for attitude change (Lee, 1995), has been identified as a key organizational process. Evidence has shown (ACSNI, 1993) that organizations with good safety records generally have effective communication systems (see Table 9.2). There appear to be a number of mechanisms for communicating safety within organizations. These are listed in Table 9.2 and are grouped under three headings: general organizational communications, health and safety communications, and management and employee processes. However, safety communication has no virtue unless it informs and changes attitudes or behaviour; Figure 12.1 highlighted 'persuasion' as a key element in 'winning hearts and minds'. The next section considers how to argue and persuade. It first considers the processes which managers can use in persuasion and then identifies the

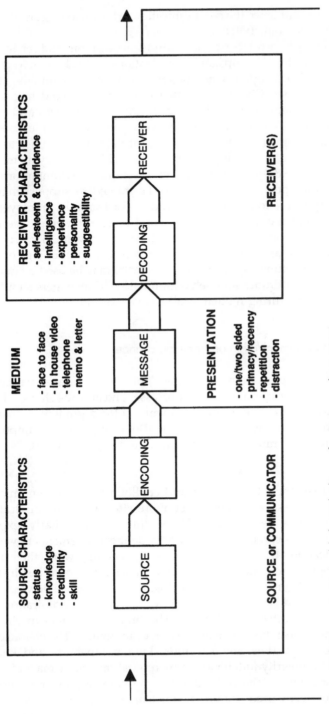

Figure 12.5 Communication model of attitude change (from S Cox and Tait, 1991)

variables that influence the communication process (see Figure 12.5 adapted from S Cox and Tait, 1991).

The most powerful form of attitude acquisition and change occurs 'naturally' within all groups and organizations and is known as conformity (Lee, 1986). What happens when people join a work system with strong and declared goals (see Chapter 9) is that their attitudes tend to converge on those that are common to that system as they interact with and learn from others. Managers can utilize the process of conformity if they nurture and reinforce a strong culture and create appropriate opportunities for learning within the work group. This process may be facilitated if key members of the group (Rogers, 1983) are encouraged to disseminate relevant information. At the same time, personnel policies can be framed to support managers in persuading people to adopt safe attitudes. For example, employees should be selected and trained appropriately and policies on employee retention, after any probationary period, should include a review of safety attitudes and behaviours, as should disciplinary codes. Conditioning, deliberately exploiting opportunities for reward (Lee, 1995) may be used by managers to encourage safe attitudes and behaviour. Basically, managers should reward safe behaviours during regular 'walkabouts' (see later).

Evidence on communication and attitude change

Figure 12.5 (adapted from S Cox and Tait, 1991) offers a diagrammatical presentation of the communication processes and attitude change. The model identifies: (1) the source of the argument; (2) the argument or message; (3) media; and (4) receiver variables. It also recognizes the importance of feedback in the communication and attitude change processes.

Source
Source variables include the status and credibility of the communicator. The argument or message is likely to be more effectively communicated if the communicator is respected and has 'high status'. Early experiments manipulated this variable by attributing identical statements to well known political leaders (Lewis, 1941). These statements were then rated for 'compellingness to action', 'social significance', 'personal inspiration' and 'author's intelligence'. There was a close correspondence between subjects' assessments and their independently determined admiration for the (experimentally attributed) leaders. The findings in this experiment have been further researched to include 'source credibility'. The characteristics of source credibility (Lee, 1995) include: known capability and knowledge, honesty and integrity (often judged by reputation), charisma and attractiveness, co-orientation (the communicator provides a good role model) and power (if correctly used). 'Sources' which are perceived as 'trustworthy' and

'attractive' are more effective in changing attitudes than those which are not (Richardson, 1977).

Message

Arguments and messages are an integral and critical part of the 'communications model' of attitude change; an interesting question arises as to whether they should be 'one sided' or 'two sided'. The following findings emerged when propaganda messages were used to change US soldiers' attitudes (see for example, Hovland *et al.*, 1949). A soldier who had received a high school education was more influenced by a two-sided communication, whereas a soldier with a poorer education was more influenced by a one-sided communication. Arguments contained in a one-sided communication were effective if the receiver's attitudes were in sympathy with that reflected in the message. An argument contained in a two-sided communication was more effective if the initial attitude of the receiver was at variance with the attitude expressed in the message. However, it has been recognized that these principles may not hold in the longer term or when counter propaganda exists. When opposing arguments (propaganda and counter propaganda) are presented, there is an order effect reflecting their primacy (first argument) or recency (last argument). There is some evidence for the superiority of primacy effect in certain circumstances. A first argument is likely to be more effective if both sides of that argument are presented by the same person, and provided the listeners are unaware that conflicting views are to be presented. If at the end of the presentation the listeners make a public commitment (see later), this is an important factor in endorsing the primacy effect.

McGuire (1969) also found that the effect of a persuasive communication can be weakened if the audience has encountered some of the arguments now being used in a previous communication which was mainly opposing the position now advocated. This is termed the inoculation effect. This argues for a *consistent* approach to safety communication.

Arguments and messages should be easily understood by those for whom they are intended and should be relevant to their needs and circumstances (Richardson, 1977). Three further aspects of the message can be considered: the first concerns the type of appeal that is being made, for example, rational or emotional (threats and fear), the second concerns the amount of attitude change that is being advocated. There is some research linking these concerns in relation to smoking cessation.

In a study by Janis and Mann (1965) women were encouraged to reduce their cigarette consumption by taking the role of female cancer patients in a role play exercise. The experimenters took the role of a physician and a series of scenes were acted out in the physician's surgery. A variety of props were employed to increase the authenticity of the role play. The women were obliged to focus on the possibility of a painful illness followed by hospitalization and an early death. Two different control groups were employed. The first engaged in cognitive role play by acting the part of debaters arguing against cigarette smoking, while the second, received exactly the same information as the role playing subjects by listening to a tape of their sessions. When interviewed two weeks later, it was found that the role play group expressed significantly less favourable attitudes to smoking than the control groups and had reduced their daily consumption of cigarettes. A follow-up, conducted for other purposes some eighteen months later (Mann and Janis, 1968), reported that those in the experimental group were still smoking less and still expressed less favourable attitudes to smoking. Comments made by the women indicated that their attitudes had been strongly influenced by the experiences in the experiment conducted eighteen months earlier.

Despite the obvious emotional impact of such an experiment, it has been fairly widely assumed since the original study by Janis and Feshbach in 1953 that the amount of attitude change after a persuasive communication was inversely related to the intensity of fear aroused by the message. Several other factors may affect this relationship, for example, the degree of change advocated and the ease with which it can be accomplished. Thus if the degree of change demanded is great and not easy to accomplish within the person's frame of reference, the fear inducing message may be rejected rather than the target attitude changed. Another factor may reflect the target for concern. Fear inducing messages are more effective if targeted on the person's loved ones or family role (see, for example, the road safety message targeted at male heavy goods vehicle drivers – 'come home safe tonight daddy').

Finally, research has shown that importance is a further factor associated with the message. People will generally not take time out to process 'unimportant' messages. This factor indicates the importance of shaping perceptions alluded to earlier in the book – safety messages must be perceived to be important.

Medium

There have been several studies designed to determine whether any one *vehicle or medium* is more effective than any other in changing attitudes. The general finding has been (Katz and Lazarsfeld, 1955; Chaiken and Eagly, 1976) that attitude change is far more likely to occur as a result of personal influence than as a result of influences originating in the media. Although personal influence may have the more direct impact on attitudes, it would be foolish to ignore the influence of the latter. Studies, such as those by Himmelweit and his colleagues (1958) have provided evidence that exposure to television does have measurable effects on the beliefs and evaluations of children concerning a wide variety of social groups and events. Indeed, an interesting conclusion arose from the work of Katz and Lazarsfeld (1955). Messages presented through the channels of newspaper, radio and television initially produced little or no change in attitudes. However, having remeasured attitudes some weeks later, the authors found significant changes had then occurred. They put forward the following explanation for these delayed shifts in opinion. To begin with, most people are affected in a very small way by what they see and hear in the press, on the radio or on the television. But they are likely to discuss these issues with others whom they know and whose opinions they value and trust (see above). It is only then that attitude change occurs to any marked degree.

However, the superiority of face-to-face contact (Lee, 1995) is unchallenged – its power far exceeds that of other media and should be used to change attitudes to workplace safety wherever practical.

Receiver characteristics
There is some evidence which deals with the effectiveness of communication in terms of the characteristics of the receiver: these include personal characteristics such as gender, age, intelligence, confidence and self-esteem. The literature on attitude change and fluoridation serves as a useful example.

In studies concerned with attitudes towards fluoridation, the pro-fluoridation message is often heard and understood but then rejected on non-rational grounds. Those who appear most opposed to fluoridation have values which, if not anti-scientific, nevertheless appear to place great emphasis upon the negative aspects of scientific achievement. Kirscht and Knutson (1961) have suggested, by way of explanation, that it is these general attitudes which set the interpretive framework for the fluoridation issue. Values precede the content of particular issues and form a primary frame of reference within which more specific frameworks are articulated. In a further study by Richardson (1977), it was found that those who were opposed to fluoridation were more likely to have a strong sense of powerlessness in which they felt unable to influence the course of their own lives. It appeared as if their negative attitudes to fluoridation were based upon a need to demonstrate some potency by applying a veto, by saying NO!

There is evidence for a curvilinear relationship between degree of change and self-esteem (Rhodes and Wood, 1992). People with very low self-esteem are less likely to process the message (or argument) because they tend to be too focused on making a good impression (Rhodes and Wood, 1992). On the other hand, those 'receivers' with high self-esteem (or confidence in themselves) are not prepared to be easily persuaded by others.

Public commitment is a powerful strengthener of attitudes and attitude change. In an experiment conducted by Hovland et al. (1949) students were asked to write an essay on their attitudes to reducing the legal voting age to eighteen years. This came after a session where they were exposed to an argument which was favourable to the idea of reducing the legal voting age. Half the group (the public commitment group) were asked to sign their essays and were told that their work would be published in full in the school newspaper the following week. The other half (the private commitment group) were not asked to sign their essays and in addition were assured that their views would remain anonymous. Both groups were then presented with an argument which was strongly in favour of retaining the then existing voting age, twenty-one years. They were invited to write a short paragraph stating their frank opinions on this issue. The results of the experiment suggested that only 25 per cent of the public commitment group shifted or changed their attitudes as a result of the second argument, while 50 per cent of the private commitment group did.

Safety management implications of the communication model

The previous sections have reviewed some of the available evidence on communication and attitude change. Such evidence suggests that each of the elements outlined in Figure 12.5 is important in the attitude change process and may be incorporated into orchestrated safety campaigns. The source of the message must be credible and have status (for example, a senior manager or an 'expert' in the particular safety issue). The content of the message must have relevance and be perceived as both important and meaningful. Face-to-face delivery is often most effective, as is the use of the media (TV or newspapers). Attention to presentation techniques reinforces the benefits of media choice. Finally, the individual characteristics of the receiver (including age, intelligence and personality) are also critical and should be taken into account.

In summary, safety attitude change campaigns should be developed in a manner which realistically address the underlying problems. They should be based on:

1 a thorough analysis of the existing attitudes, their architecture and the key 'driving' forces;
2 appropriately designed and focused messages that are strategically targeted;
3 preliminary testing to see if they have a reasonable change of succeeding (Lee, 1987); and
4 evidence of the overall success, i.e. impact on safety performance. For example, Saarela et al. (1989) showed the lack of effect of a safety poster campaign upon accident rates. The campaign was well received but had no measurable effect on individuals.

Changing Behaviour

The previous sections have dealt with the communication model of attitude change, the following sections deal with the alternative approach based on behavioural change. Behaviour modification techniques have been used for two purposes – first, as a vehicle for attitude change and, second, as a direct vehicle for changing behaviour. Effective behaviour modification deals with carefully specified and discrete behaviours and, in the first case, it is assumed that changing particular behaviours will promote more general attitude change which, in turn, will underpin an apparent 'halo' effect and a more widespread effect on behaviour.

The techniques which are used to shape or modify behaviour have been generally derived from early learning theory (Skinner, 1938). The German psychologist Burkhardt has applied such behaviour modification techniques to workplace safety (Burkhardt and Schneider, 1987). His work has focused on areas of high accident frequency and is based on the following principles:

1 actively reinforcing safe behaviour patterns; and
2 extinguishing unsafe work behaviour.

The operational steps of the method are described in Table 12.3 together with suggested implementation techniques. It is argued (Burkhardt and Schneider, 1987) that the key to a successful programme is the systematic application of the following behavioural guidelines.

1 *Employees should be rewarded for behaving safely.* This requires that safe behaviour should be clearly defined and obvious in any workplace and for any procedure. Whenever possible employees should be treated individually and not as a group.
2 *Employees should be satisfied with the nature of the rewards given.* The rewards should be proportional to the effort expended and the level of perform-ance. Employees should believe that they can perform at the specified level, and should be given regular feedback on how well they are doing. The allocation of rewards should be clear and understood by all employees.
3 *Rewards should be plentiful enough to be an important part of day-to-day working but not too plentiful to become expected or routine.* The reward principle should be consistently applied, and employees should be rewarded as near in time to the act of safe behaviour as possible. It may not be desirable to reward every act of safe behaviour but to present reward on some predetermined schedule (say once every ten accumulated acts). This is referred to as partial reinforcement.
4 *The act of reward should be consistent with the prevailing management style,* and the reward should be valued by the individual, his/her colleagues and the organization. Rewards should compare both within and across organizations and divisions.

While safe behaviour is rewarded, it is important to ensure that unsafe behaviour is not. By avoiding rewarding unsafe behaviour and by replacing

Table 12.3 Burkhardt's behaviour modification techniques (Burkhardt and Schneider, 1987)

Operational step	Methodology
1 Identification of accident and hazard concentrations	Definition of accident concentrations on the basis of accident statistics Subsequent analysis of accident-concentrations according to causation, body part, etc. Measurement of the degree of awareness of typical risks by supervisors and co-workers
2 Revising the behavioural rules which have to be observed in order to prevent the accidents	Revision of the existing rules; if necessary, completion or alteration of the rules. Ranking the order of their importance Investigating the degree of realization rules Selecting a final set of rules according to importance and degree of realization as a basis for further working steps
3 Developing a plan of reinforcers and extinguishers to modify the working behaviour	Finding measures which are able to positively reinforce the safe behaviour Looking for measures which can reduce the disadvantages of safe behaviour Considering the measures which will reinforce the disadvantages of unsafe behaviour Gaining measures which reduce the advantages of unsafe behaviour Analysing justifications of unsafe behaviour and defining arguments to work them off
4 Realization of the plan of reinforcers and extinguishers	Dividing the proposals into training procedures and long term application techniques Defining the target groups, in most cases being: – the group of supervisors in the chosen area – the people working in the area of accident concentrations Realization of the measures
5 Follow-up study for measuring the effects of the method performed	Controlling the degree of realization of the above plan Multi-moment studies for evaluating the degree of rule observance Analysing the development of accidents before, during and after the study

it in competition with safe behaviour, it should extinguish. Petersen (1989) has cited a number of studies that have used reinforcement principles to improve work practices and has recorded an associated reduction in workplace injuries.

This emphasis on reinforcing safe behaviours has been applied in safety interventions in several UK organizations (see, for example, Sutherland and Martin (1995) in a chemical company, Robertson *et al.* (1995) in the construction industry, and S Cox and Vassie (1995) in manufacturing). It is also at the heart of Krause *et al.*'s (1990) approach which is described in their book on *The Behaviour Based Safety Process.*

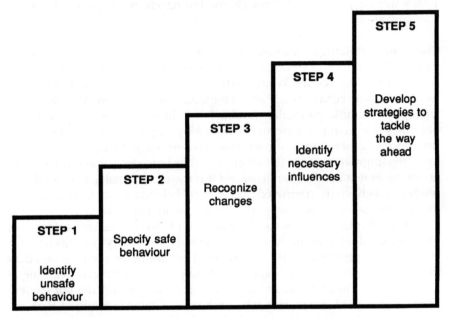

Figure 12.6 Five steps in safe and healthy behaviours (from S Cox and Vassie, 1995)

The first author has proposed a five-step behaviour based safety programme (see Figure 12.6) aimed at empowered work teams which builds upon the principles outlined above (S Cox and Vassie, 1995). It focuses on the identification, monitoring and reinforcement of safe behaviours, and involves workers challenging any unsafe behaviours. A systematic approach is required which is supported by expert facilitators. A toolkit (S Cox and Vassie, 1995) has been produced both to facilitate the process and to enable work groups to develop their own approaches to behavioural interventions.

Evidence on the success of behavioural programmes (Sutherland and Martin, 1995; Robertson *et al.*, 1995; Krause *et al.*, 1990) would seem to suggest that they are extremely successful in reducing workplace accidents.

However, there are several principles which need to be adopted early in any such programme to ensure success:

1 behavioural programmes should not be considered in isolation of other initiatives;
2 they should be designed to fit the prevailing culture (particularly in terms of appropriate reward systems) and have the commitment and 'overt' support of senior management;
3 behavioural programmes require expert and patient facilitation and do not necessarily produce immediate improvement; and finally
4 they are only sustained through continuing efforts at all levels of the organization.

One of the continuing debates on the impact of behavioural programmes is the nature of 'incentives' or rewards. Several researchers (see, for example, Sulzer-Azaroff, 1982; McAfee and Winn, 1989) have reviewed the use of incentives in relation to safety behaviours. Many of the studies examined in the reviews had attempted to link particular feedback on incentives with behavioural outcomes (for example, percentage of employers performing the job safely, frequency of observations of particular hazards, etc.). All the studies found *short term* improvements in at least one variable; however, some studies found no improvement in other variables and it is therefore possible that, while the rewarded behaviour improved, other safe behaviours deteriorated. The improvements were not always sustained in the longer term.

Earlier chapters of the book have emphasized the importance of organizational learning in the development of good safety cultures (see, for example, Chapters 9 and 10). Behavioural programmes offer an excellent vehicle for organizations to express and support organizational learning at all levels. The initial focus on accident concentration and the subsequent analysis of hazard–harm–risk relationships facilitates learning at both the individual and organizational levels. Equally, the decisions on what are safe behaviours and practices confirm the importance of such learning and reinforce the safety messages (ACSNI, 1993). This approach may also ensure that safe behaviours are sustained over longer periods of time.

Training

While two contrasting approaches to attitude change have been described in the previous sections, that contrast should not be taken to imply that they are incompatible in all respects. That is not the case and, in practice, most effective programmes involve both verbal or written communication, and the shaping and reinforcement of the desired response to that communication. This blend of approaches is obvious in most training programmes.

Training has been used as a vehicle for attitude change, and there is an emergent literature dealing with this issue (see, for example, Patrick, 1992). The first author has described a training intervention within a European industrial gas company (S Cox, 1988, 1995) which builds on the earlier example of attitude measurement. The intervention focused on two issues – company commitment and leadership for safety. Five groups of supervisors (approximately fifteen to twenty per group) attended the training work-shops which were participative in nature. In the leadership exercises supervisors were asked to assess the workplace risks associated with eight potentially hazardous situations described in vignettes; for example, one vignette described a situation where a worker was raised on the forks of a fork-lift truck to change a lightbulb. The supervisors were then required to manage the situation. They were further tasked (through the use of video playback) with exploring appropriate influencing skills to ensure that workers adopted safe working behaviour in relation to the scenarios. The supervisors whose subjective assessments matched the 'objective' risks were better able to develop effective influencing skills (this may reflect their own attitudes to safety). Influencing skills are essential for managers and team leaders in order to secure effective safety.

The training intervention can also be reviewed in terms of Ajzen's (1991) model (see Figure 12.3 earlier).

1 The programme was designed to address attitudes (and their relevance) and perceptions of risk, particularly in relation to the company's safety philosophy. These in turn were linked to safe behaviours (much of the influencing was face-to-face).
2 It directly challenged individual behaviours and explored (through role play) possible views on 'perceived control'.
3 Social norms were developed within teams and groups of trainees which fostered positive attitudes.

Finally, the programme was carried out in a 'safe' environment which was supportive of the learning process.

A follow-up study (Cox, 1988) on the attitudes of the supervisory group was used to evaluate the success of the intervention. The evaluation data described an interesting change in the supervisors' attitudinal framework. The training workshop was designed to improve understanding and belief in the company's safety philosophy and commitment. This it did, the data clearly demonstrate that the supervisors better understood the philosophy and subsequently reflected it in their attitudes. They expressed a stronger belief in the company and its commitment. However, they did not appear to develop more positive attitudes to safety in general and were sceptical whether their colleagues believed in the company's safety commitment.

Several interesting things might have been happening around these latter observations. First, their attitudes to general safety were initially more positive than those held in relation to the company's philosophy (see Figure 12.4). Second, the company had been under considerable commercial pressure preceding the training. Finally, as the supervisors became more aware of their safety management role through training they could have observed more examples of unsafe behaviours in their immediate work groups. This may have accounted for the shift in scepticism.

The final sections of this chapter move on from issues of attitude and behavioural change to consider some of the key principles of learning in the context of training for safety.

Learning principles: the basis for training

Learning is a complex function (see Chapter 7) which involves both cognitive (mental) activity and non-verbal and verbal behaviours. Kolb (1984) has described the process in terms of a learning cycle (see Figure 12.7). This cycle (Kolb, 1984) describes how people learn by testing their understandings (i.e. personal theories) against their own experiences and modify their understandings as a result. Basically, he believes that *individual* experiences are the basis for learning. Kolb's model of learning is thus an example (at the individual level) of a simple feedback system. However, it is

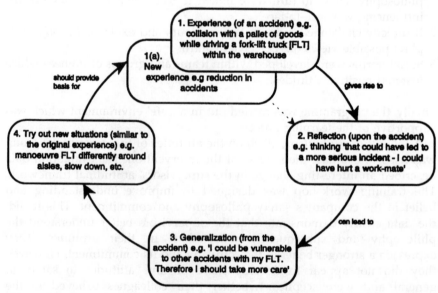

Figure 12.7 Kolb's learning cycle: a safety example (adapted from Glendon and McKenna, 1995)

arguably a difficult process to sanction in certain workplace conditions, for example, in extremely high risk situations the person would be severely harmed before they had a chance to reflect. The key to effective training is thus a balance between self-discovery and appropriate 'direction' and supervision. Glendon and McKenna (1995) have illustrated Kolb's (1984) cycle in relation to a road safety example.

Figure 12.7 illustrates a similar example in relation to workplace safety. This model also illustrates the interplay between behaviour and attitudes discussed in earlier sections. It also highlights the possibility of 'simulating' experiences in a safe learning environment, a practice which is fully exploited in many training situations (particularly in the aerospace industry). The first author (S Cox, 1987) has attempted an overview of current needs for safety training which considers the nature of organizational practices in relation to workplace safety.

Honey and Mumford (1986) have developed Kolb's (1984) ideas and further postulated that individuals have preferred styles of learning and that these preferences could be related to different stages of the learning cycle. Thus some people prefer to learn though active experiences whereas others prefer to spend time on reflection. Honey and Mumford's (1986) learning studies showed that the most effective 'learners' focused on all stages of the cycle.

Although individuals differ in their preferred learning styles and these may have some bearing upon training methods, there are a number of general issues which need to be considered in the design and execution of effective training, including training methods, task complexity, performance requirements, etc.

Table 12.4 (adapted from S Cox et al., 1995) illustrates various training techniques. The chosen method must relate to the exact nature of the particular task (for example, whether it is a simple or complex task) and the particular performance requirements (for example, is the person experiencing familiarization or competence training). Rasmussen's (1980) SRK model of task performance (see Chapter 7) describes three separate levels of performance including skills, rules and knowledge. The training requirements for these three levels are as follows:

1 Skills – repetition of the task sufficient times, with trial and error, in order to develop the skill as it should be performed. This may be achieved by effective on-the-job training.
2 Rules – this involves more structured teaching of the sequence of tasks so that the complete sequence is perfected. Checklists or flowcharts (see Chapter 7) may be useful as aide-memoires and safe working procedures should support effective learning
3 Knowledge – this level of performance operates at the problem solving level and can be supported through the use of a number of case studies,

Table 12.4 Training techniques (adapted from S Cox *et al.*, 1995)

Learning type	Examples
Lectures	External professional meetings
Discovery learning	Workshops
	Simulation/role-play
	Case studies/projects/syndicates
	Brainstorming/lateral thinking
Open learning	Textbooks/handouts
	Databanks
	Computer based training (CBT)
	Video
	Programmed instruction
Tutorials/supervision	Informal learning
	Via media, for example TV
	Discussion/debate
	Role models
	Toolbox talks
Practical training	Placements
	Learning by doing

brainstorming and lateral thinking. Training at this level is usually focused on a variety of the methods in Table 12.4 and is subject to vigorous validation. The key to effective performance at this level is that the trainee recognizes the problem as a knowledge-based rather than a rule-based problem.

Finally, Glendon and McKenna (1995) have stressed the importance (for safety) of individuals learning appropriately from experiences. This is particularly important for the development of positive attitudes and safe behaviours (see Figure 12.3 earlier). For example, training courses which build on trainees' 'near miss' (van der Schaaf *et al.*, 1991) experiences through discussion can be an invaluable way of increasing awareness of risk and improving subjective norms. They can also affect individual perceptions of behaviour control which feed directly into individual behaviours for safety.

Summary

This chapter has focused on attitudes and behaviour and their importance for safety management. The relationship between attitudes and behaviours is generally acknowledged to be an issue which is central to workplace safety. Whereas several theories (see earlier) have explored the relationships between attitudes and behaviour and have established links, the direction of the linkage is still unclear. Despite this there is much interest within organizations in developing techniques for attitude change and for modifying behaviours. Attitude change models tend to be based on the process of communication and are often treated separately from behavioural interventions. Training combines communication and attitude change with behaviour modification and supports the development of appropriate skills, procedures and knowledge. However, these positive training outcomes are not always guaranteed. Training will not work unless it is well planned, executed and is evaluated. Furthermore, the evaluation data should be fed back to support further development.

The following key points provide a practical summary.

1 Attitudes are an important factor in support of continuing improvements in workplace safety. They are the basis of an effective safety culture.
2 Attitudes and behaviours are related (see earlier) although the nature of the linkage is complex.
3 Contemporary theories would suggest that the following need to be taken account of in the process of an attitude change for safety:
   (i)   individual workers (at all levels) need to have formed an attitude to a particular aspect of safety and be *aware* of the relevance of their attitude for safe behaviour;
   (ii) the organizational context and prevailing social environment should create norms which support and reinforce such attitudes;
   (iii) individual perceptions of control actions need to be realistically framed.
4 A well developed technology exists for the elicitation and measurement of safety attitudes. The results can be used for monitoring the 'health' of a safety culture and for designing and implementing 'therapeutic' interventions.
5 Attitude change programmes have been based on (a) communication models and (b) behavioural programmes.
6 Training combines communication for attitude change with behavioural modification and can be used as an effective vehicle in support of managing people and their attitudes to safety.

References

Advisory Committee for Safety in Nuclear Installations (ACSNI) (1993). *Third report: 'Organising for Safety' of the Human Factors Study Group of ACSNI*. Sudbury, Suffolk: HSE Books.

Ajzen, I. (1991). 'The theory of planned behaviour'. *Organizational Behaviour and Human Decision Processes*, **50**, 179–211.

Burkhardt, F. and Schneider, B. (1987). 'A five-step method to modify behaviour in accident/incident concentrations'. In *Successful Accident Prevention Programmes, Arbete och Hala* (E. Menckel, ed), **32**, 35–42.

Chaiken, S. and Eagly, A.H. (1976). 'Communication modality as a determinant of message persuasiveness and message comprehensibility'. *Journal of Personality and Social Psychology*, **34**, 605–614.

Cheyne, A. and Cox, S. (1994). 'A comparison of employee attitudes to safety'. In *Proceedings of IVth Annual Conference on Safety and Well-Being at Work, November 1994* (A. Cheyne, S. Cox and K. Irving, eds). pp. 57–64, Loughborough: Loughborough University of Technology.

Cox, S.J. (1987). 'Safety training: an overview of current needs'. *Work and Stress*, **1**(1), 67–71.

Cox, S.J. (1988). *Attitudes to Safety*. Unpublished M.Phil. Thesis, Nottingham: University of Nottingham.

Cox, S.J. (1995). 'Risk perception and its influence on safety behaviour'. Paper presented to *Understanding Risk Perception*, 2 February, The Robert Gordon University, Aberdeen.

Cox, S.J. and Cox, T. (1991). 'The structure of employee attitudes to safety – a European example'. *Work and Stress*, **5**(2), 93–106.

Cox, S.J. and Tait, N.R.S. (1991). *Reliability, Safety and Risk Management*. London: Butterworth-Heinemann.

Cox, S.J. and Vassie, L.H. (1995). *Behavioural Safety Toolkit*. Loughborough: Loughborough University of Technology.

Cox, S.J., Janes, W., Walker, D. and Wenham, D. (1995). *Office Health and Safety Handbook*. Croydon, Surrey: Tolley Publishing Company.

Ferguson, E. and Cox, T. (1993). 'Exploratory factor analysis: a user's guide'. *International Journal of Selection and Assessment*, **1**(2), 84–94.

Festinger, L. (1957). *A Theory of Cognitive Dissonance*. Evanston, IL: Ron Peterson.

Glendon, A.I. and Hale, A.R. (1984). 'Taking stock of site safe '83'. *Health and Safety at Work*, **6**(8), 19–21.

Glendon, A.I. and McKenna, E.F. (1995). *Human Safety and Risk Management*. London: Chapman and Hall.

Health and Safety Executive (HSE) (1993). *Attitudes towards noise as an occupational hazard*. HSE Contract Research Report No. 55/1993. London: HMSO.

Himmelweit, H.T., Oppenheim, A.N. and Vince, P. (1958). *Television and the Child*. London: Oxford University Press.

Honey, P. and Mumford, A. (1986). *The Manual of Learning Styles*. Maidenhead: P. Honey.

Hovland, C.I., Lumsdaine, A.A. and Sheffield, F.D. (1949). *Experiments in Mass Communication*. Princeton, NJ: Princeton University Press.

Janis, I.L. and Feshbach, K.S. (1953). 'Effects of fear arousing communication'. *Journal of Abnormal and Social Psychology*, **48**, 78–92.

Janis, I.L. and Mann, L. (1965). 'Effectiveness of emotional role playing in modifying smoking habits and attitudes'. *Journal of Experimental Research in Personality*, **1**, 84–90.

Katz, E. and Lazarsfeld, P.F. (1955). *Personal Influence*. Chicago: Free Press.

Kelman, H.C. (1961). 'Processes of opinion change'. *Public Opinion Quarterly*, **25**, 57–78.

Kirscht, J.P. and Knutson, A.L. (1961) 'Science and fluoridation: an attitude study'. *Journal of Social Issues*, **17**, 37–44.

Kleinke, C.L. (1984). 'Two models for conceptualising the attitude-behaviour relationship'. *Human Relations*, **37**(4), 333–350.

Kolb, D.A. (1984). *Experiential Learning*. Englewood Cliffs, NJ: Prentice-Hall.

Krause, T.R., Hidley, J.H. and Hodson, S.J. (1990). *The Behaviour-Based Safety Process: Managing Involvement for an Injury-Free Culture*. New York: Van Nostrand Reinhold.

Lee, T.R. (1986). 'Effective communication of information about chemical hazards'. *The Science of the Total Environment*, **51**, 149–183.

Lee, T.R. (1987). *Action on attitudes to risk in Department of Health and Social Security strategies for accident prevention*. Report of a Colloquium for the Medical Royal Colleges of the UK on 26 March 1987 under the auspices of Medical Commission on Accident Prevention. London: HMSO.

Lee, T.R. (1993). 'Psychological aspects of safety in the nuclear industry'. The Second Offshore Installation Management Conference, 'Managing Offshore Safety', 29 April, Aberdeen.

Lee, T.R. (1995). 'The role of attitudes in the safety culture and how to change them'. Paper presented to the Conference on *Understanding Risk Perception*, February, The Robert Gordon University, Aberdeen.

Lewis, H.B. (1941). 'Studies in the principles of judgements and attitudes: IV. The operation of "prestige suggestion"'. *Journal of Social Psychology*, **14**, 229–256.

Likert, R. (1932). 'A technique for the measurement of attitudes'. *Archives of Psychology*, **140**, 52.

Mann, L. and Janis, I.L. (1968). 'A follow-up study on the long-term effects of emotional role playing'. *Journal of Personality and Social Psychology*, **8**, 339–342.

McAfee, R.B. and Winn, A.R. (1989). 'The use of incentives/feedback to enhance workplace safety: a critique of the literature'. *Journal of Safety Research*, **20**(1), 7–19.

McGuire, W.J. (1969). 'The nature of attitudes and attitude change'. In *Handbook of Social Psychology*, Vol. 13, 2nd edn (G. Lindsey and E. Aronson, eds), New York: Addison-Wesley.

Patrick, J. (1992). *Training: Research and Practice*. London: Academic Press.

Petersen, D. (1989). *Safe Behaviour Reinforcement*. New York: Aloray.

Purdham, J.T. (1984). *A Review of The Literature on Attitudes and Roles and Their Effects on Safety in The Workplace*. Hamilton, Ontario, Canada: Canadian Centre for Occupational Health and Safety.

Rajecki, D.W. (1990). *Attitudes*. Sunderland, MA: Sinauer Associates.

Rasmussen, J. (1980). 'What can be learned from human error reports?' In *Changes in Working Life* (K.D. Duncan, M. Gruneberg and D. Wallis, eds), London: Wiley.

Rhodes, N. and Wood, W. (1992). 'Self esteem and intelligence affect influenceability: the mediating role of message reception'. *Psychological Bulletin*, **111**, 156–171.

Richardson, A. (1977). 'Attitudes'. In: *Introductory Psychology* (J.C. Coleman, ed), London: Routledge and Kegan Paul.

Robertson, I.T., Duff, A.R., Marsh, T.W. *et al.* (1995). 'Reducing unsafe behaviours –how to get rid of those stubborn accident statistics'. International Safety and Health at Work Conference, Olympia, London.

Rogers, E.M. (1983). *Diffusion of Innovations*, 3rd edn. New York: Free Press.

Saarela, K.L., Saari, J. and Aaltonen, M. (1989). 'The effects of an informational safety campaign in the shipbuilding industry'. *Journal of Occupational Accidents*, **10**, 255–266.

Skinner, B.F. (1938). *The Behaviour of Organisms*. New York: Appleton Century Crofts.

Snyder, M. (1979). 'Self-monitoring processes'. *Advances in Experimental Social Psychology*, **12**, 85–128 (L. Berkowitz, ed). New York: Academic Press.

Snyder, M. and Kendzierski, D. (1982). 'Acting on one's attitudes: procedures for linking attitude and behaviour'. *Journal of Experimental Social Psychology*, **18**, 165–183.

Sulzer-Azaroff, B. (1982). 'Behavioural approaches to occupational health and safety'. In *Handbook of Organizational Management* (L.W. Frederiksen, ed.) 505–538, New York: Wiley.

Sutherland, V. and Martin, P. (1995). 'Changing behaviour, improving safety: a behavioural approach'. *Health and Safety Bulletin*, **229** (January).

Van der Schaaf, T.W., Lucas, D.A. and Hale, A.R. (eds) (1991). *Near Miss Reporting as a Safety Tool*. Oxford: Butterworth Heinemann.

The way ahead

The authors have argued in this book that the effective management of safety is important for all organizations, and have endeavoured to provide some insight into how that might be best achieved. In doing so, they have refined the basic argument in two ways: first, they have argued, on the basis of complexity theory, that no other goals are more important than safety, and, second, they have argued, on the basis of evidence provided by the UK HSE, that organizations that are successful in managing safety are often successful with respect to other of their goals. These two arguments are linked and should provide a powerful insight into how organizations should position the challenge of safety management in the wider organizational context.

The authors' approach to meeting this challenge has formed the framework for this book. Essentially, what is presented is a systems-based view of the interaction between a wide range of psychological, social and organizational factors involved in the effective management of safety. The framework has combined three different perspectives:

1 an overarching systems theory perspective, which focuses, in part, on
2 the person in their job in their organization, and which
3 treats the person as an active information processing system.

The title of this book – *Safety, Systems and People* – was chosen to capture this interplay of the ideas.

Systems theory was used to provide the language and thinking which bound the framework together and to provide the scaffolding for its other aspects. The book's exploration of the fundamentals of systems theory was developed in a number of ways, covering new topics such as complexity theory, the theory of organizational health, and the use of metaphors. At the same time, much was made of sociotechnical theory in the application of systems thinking to the analysis of organizations. These different topics and theories were introduced in the early sections of the book, and woven together to provide a basis for the more practical information presented in the final section. The systems approach was exploited in the chapters dealing first with the organization, and then the job and finally the person. The human information processing model of the person (at work) was suggested as the most adequate in this context, although challenges to this model from computational psychology and neuroscience were recognized.

Taken together, these different domains of knowledge map out the contemporary psychology of safety, which deliberately takes the reader away from issues solely relating to individual differences and personality, and an exclusive and near clinical emphasis on what the individual thinks and does in making errors. That approach implicitly attributes blame for safety failures to the individual, and blindly ignores their technical, social and organizational antecedents. Threatening and cajoling individuals to behave more safely, punishing them for not doing so and rewarding them for succeeding can change the profile of their behaviour, and likely as not change the way those individuals think about the organization, but such strategies on their own will not necessarily improve overall safety performance. In particular, the application of punitive strategies may give rise to resentment and reduce the individual's commitment to the organization and its goals including safety. Such strategies may only work when there is a strong physical presence of management. This argument is made clear in the latter part of the book, when the more theoretical information is translated into practical advice on organizational and management actions. A truly integrated approach is spelt out there.

In these latter chapters, the book considers the organization's goals and objectives, strategies and tactics for effectively managing safety, and it explores what needs to be done at the levels of the organization, job and individual. Throughout it argues for a comprehensive approach which integrates thinking about the hardware and software with that about the human factor represented in the person in their job in their organization.

Managing safety in organizations, at all levels, is about systems control. There are two aspects to such control: control by design, and control through ongoing management action. It was argued that, in the present context, these two aspects are strongly related as managers have a responsibility not only for the actions which build into the process of managing but also for the design of the organization, its work, work equipment and work environments. A balance is struck in the book between ensuring safety through design and through the ongoing process of managing.

Early in these chapters, it is argued that managers have to recognize that there are no such things as certainty and perfection in the imperfect and ever changing world of organizations. This means that the risk of safety failures will never be reduced to zero, nor is it possible to design perfect systems. In the management of safety, as with all other aspects of organizations, good enough solutions are good enough. However, the corollary to this is that management has to be flexible and adaptable to that changing world: what is a good enough solution today might not be tomorrow. Hence there is a need for continual review and monitoring, and the availability of good information on safety is critical to the effective management of that function. It has also been argued that such information should be widely available, and not simply held by management. It should be pushed down the organizational hierarchy as far as is possible and sensible, and decisions involving safety performance should also be subject to this principle of subsidiarity.

According to the authors' line of argument, the successful organization is also a safe and healthy organization, and the safe and healthy organization is likely to be a successful one. Such organizations are productive, flexible and creative; they evolve and adapt. They look after their staff because in doing so they look after themselves. Such organizations are fit for their declared purpose, and they are fit to face the future. What does the future hold?

The future

The process of predicting future trends in any area of organizational endeavour, such as safety, is fraught with difficulties. First, it assumes that there are historic trends which underpin and make possible such predictions. This, in turn, implies that events are linked over time and are not more or less random: one offs. Second, it assumes that these trends may be generalized both within and between sectors, industries and organizations so that conclusions can be drawn which are at a sufficiently high level to be of general interest. Finally, organizations and their safety management systems operate within their wider environments (see Chapter 4) and are thus affected by changes in such environments. The prediction of changes in the latter is no small feat in itself (see later). The accuracy of these predictions establish the limits to predicting future trends in safety. Despite these difficulties and accepting that the future predictions may be flawed, the process of looking ahead has value. Thinking about what may happen should force a better understanding of the present. The adequacy of all models and theories is dependent on them being capable of prediction as well as description, and stretching the time base for that prediction can be a useful test of their strength.

John Rimmington retired on 30 June 1995 as the Director General of the UK HSE, a post he had held for eleven years. In a valedictory summary of health and safety since the HASAWA 1974 he identified six principal trends:

1 A rise in concern for public safety as opposed to concern for issues which only affect workers.
2 Changes in the structure of industry 'away from the once dominant people-intensive and unionized political platforms towards small scale, mobile units often under contract and under pressure'. This quote captures the trend towards small businesses.
3 Internationalization, including the impact of major accidents and the single market.
4 An insidious rise of liability, certain risks are now virtually uninsurable, and the losses on the employers' liability market, mainly from 'delayed' health risks, are leading to a restriction in the cover afforded.
5 A move towards a greater concern for 'health' compared to 'safety' risks for example those harms associated with exposure to chronic hazards are now assuming greater importance as traditional safety hazards have been better controlled.
6 An increased emphasis on the human factor in the causation of accidents and the management of safety, using the same techniques as are applied to other aspects of company performance.

More generally, Rimmington believes that all these trends can be rationalized in terms of a significant change in the way organizations interface with the public. He specifically commented on:

> Public concern about the effects of technology on life and the environment, and the preventative and compensatory approaches needed to satisfy this while allowing necessary technical progress and wealth creation.

These are the issues that Rimmington has drawn attention to. His list of future developments is both informed and informative. However, it can be reasonably extended by exploring possible developments in the wider environment for work. The present authors offer six supplementary issues for consideration. They relate to the changes that are known to be occurring in the nature of organizations and the very structure of work; all will impact on current thinking about the management and practice of safety. The issues are:

1 The increasing globalization of industry and commerce, which is much discussed, set against an increasing awareness of nationality and an associated strengthening of the local defence of national interests, culture and environments.
2 The development of new organizational structures and forms of management with the evolution of more extreme structures and management practices reflecting a growing divide between work which is of a relatively

low knowledge and skill requirement and that which is dependent on high level expertise and creativity (organizations supporting the former may become increasingly authoritarian, centralized and bureaucratic while the latter may become more democratic, decentralized and flexible).

3 A continuing shift in the nature of work from the full-time to the part-time and from long term tenure to temporary employment with the increasing participation of women in the workforce and the increasing salience of psychosocial, rather than physical, demands and hazards.

4 An increasing use of information and communications technologies creating smaller work teams, and more isolated workplaces with teleworking and the electronic cottage being an increasing reality.

5 An increasing size gap between the large multinational organizations and small and medium sized businesses with the latter becoming increasingly important for local, if not national, economies.

6 An increasing reappraisal of the role of work in the structure of peoples' lives, and the role of people in the work process which will lead to the acceptance of a high level of structural unemployment and the challenge that this brings to social policies.

These six issues add up to a major challenge to the way in which we think about people, work and the nature of our society, and to our thinking about what can be achieved in relation to the management of safety at work. Consider the example of the teleworker.

A teleworker is employed as a sub-contractor on piece-rate, and works at home with his or her family around them. They work on their own with their own equipment set up in what would otherwise be the spare bedroom. The equipment is not fully adjustable, and the children play with it when it is not being used for work. The teleworker's performance is electronically monitored. The basic question is how can their safety be protected? The subsidiary questions are: 'from what and from who?' and 'with how much interference from outside into the privacy of family life?' A wider question is 'who cares enough to work these questions through?' 'Where will the political clout emerge from to force society to care?'

Large multinational companies may need to think more carefully about whether and how they develop crossnational procedures and standards, and whether they should move towards the more equitable treatment of workers in different countries. Various forces will be brought to bear on them in deciding on this issue, including the interaction between their organizational 'rules' and the health and safety laws of the countries 'housing' their offices and operations, the economic imperative on both those countries and on the organization, and local community action in defence of the local environment, culture and interests.

At the same time, organizations will have to think through the positioning and nature of the safety function especially where they are moving towards greater decentralization with increased use of information and communications technology, and short term and/or piece-rate employment contracts. There will be a particularly strong and difficult challenge both where the workplace is the home and the worker is the teleworker, and where the organization is small and the workers are all sub-contracted. In the first case, the need to manage safety will begin to interface with issues of privacy. In both cases, the move towards the democratization of safety management may be countered by the need to impose ever stricter systems control over individuals as other pressures on them increase. The interplay between productivity and financial reward, currently faced at the organizational level, will be pushed down to the individual level in both cases.

Finally, if work loses its centrality, and relatively fewer people are engaged in a lifetime of work, then the importance of occupational safety will diminish as a national political argument. The emphasis will move from occupational health and safety to public and environmental health and safety. This shift is already occurring. Where an interest in work-related safety remains, it may be focused more on the impact of work environments and systems on customer and client groups than on workers.

It must be clear to all those involved in the management of safety, and who think about the future of work, that major changes are afoot. Given that we have not yet got things right in the current context, we face twin challenges: to force improvements in the way we do things now, and to develop new ways of doing those things for the immediate future. Can these twin objectives be achieved together? Can present practice evolve seamlessly into future practice or is a major revolution in thinking about to happen? Can occupational safety survive these changes?

The authors' offer these questions, and leave the reader the challenge of thinking through the answers.

Appendix 1
Safety Management Systems (SMS) audit (sample page)

1 Policy

Management

1a Is the health and safety policy articulated (in line with other business considerations) and in place?
If the answer is YES, where is the policy displayed and when was it last reviewed?

1b Further areas for consideration:
    (i)   How is the health and safety policy communicated?
    (ii) Is the health and safety policy understood?
    (iii) Is health and safety considered in each business decision?
    (iv) Is a regular review made of the site policy against actual achievements?
    (v) What arrangements are there for updating the policy?

Workforce

1c Is there a health and safety policy in place?
If the answer is YES, where can you obtain access to it and when did you last read it?

1d Further areas for consideration:
    (i)   How is the health and safety policy communicated to the workforce?
    (ii) Do you feel that the health and safety policy is appropriate to the work undertaken?
    (iii) How does the health and safety policy impinge on your work area?
    (iv) How do you feel that you can influence health and safety policy?

Appendix 2 Safety attitude questionnaire (sample)

Please read each statement and, using the scales provided, say to what extent you agree or disagree with it; please circle the appropriate reply.

	I strongly agree	I agree	I neither agree nor disagree	I disagree	I strongly disagree	I do not understand the statement/I have no opinion
1 Safety is not only the responsibility of the company but also of the individual worker	1	2	3	4	5	9
2 Sometimes it is necessary to take risks to get a job done	1	2	3	4	5	9
3 Safety works until we are busy, then other things can take priority	1	2	3	4	5	9
4 It is important for everyone to have regular safety updates	1	2	3	4	5	9
5 I should encourage my colleagues to work safely	1	2	3	4	5	9
6 There is no point in reporting a near miss	1	2	3	4	5	9
7 Not all accidents are preventable	1	2	3	4	5	9
8 Accidents only happen to other people	1	2	3	4	5	9
9 People who work to procedures will always be safe	1	2	3	4	5	9
10 After an accident the company is more concerned with apportioning blame than future prevention	1	2	3	4	5	9
11 My manager/supervisor often reminds me of safe working practices	1	2	3	4	5	9

Index